PICAFLORES
The Nerve Endings of GOD

By

Matthew J. Pallamary

Mystic Ink Publishing

Mystic Ink Publishing
San Diego, CA
mysticinkpublishing.com

ISBN: 978-0-9986809-7-2 (sc)

Library of Congress Control Number: 2021923514
Mystic Ink Publishing, San Diego, CA

Book Jacket and Page Design: Matthew J. Pallamary / San Diego CA
Cover artwork: Matthew J. Pallamary / San Diego CA
Author's Photograph: Matthew J. Pallamary — Gibbs Photo / Malibu CA

Acknowledgements

The author would like to thank Rob and Kim Gubala whose support made this book possible.

TABLE OF CONTENTS

LOVE AT FIRST FLIGHT

My magical, otherworldly, intimate connection with the spirit of Picaflor comes from two decades of visionary journeys experienced within the context of shamanic plant diets in the Peruvian Amazon where it was first gifted to me by the loving spirit of Condor. I relate to the Spanish name of Picaflor which means to bite or sting flower because this is the common reference used by the Shipibo Indians, mestizos, and other groups I have worked with in Peru.

Going back further in history, the Asháninka are indigenous people living in the rainforests of Peru and in the State of Acre, Brazil. Their ancestral lands are in the forests of Junín, Pasco, Huánuco and part of Ucayali in Peru. Tonkiri is their word for hummingbird. Colibri is the common name used in Mexico which comes from the French Colibri that has its roots in the Latin Colubris. In Portuguese, hummingbirds are known as Beija-Flor, which means flower kisser and in the Dominican Republic people call them suma flor, buzzing flower. In other Spanish uses they are known as Chupaflor or Joyas Voladoras, meaning flower-sucker, or Flying Jewels, and in the Caribbean, they are commonly referred to as "El Zunzun" which translates as going fast, referring to their quick movements.

In my visionary journeys I have been blessed to experience reality filtered through a shamanic world view where *everything* is energy, and from this magical point of view the word spirit and energy mean the same thing. If you look at the world amplified through the lens of this timeless supernatural perspective, you will gain insight into the profound ways that plants, animals, and insects communicate with us and each other.

On an olfactory level, plants call to us from a distance by their varied aromas that form the basis for many scents and perfumes that are attractive to humans, often to the point of stimulating a resonance between lovers. Additionally they are used as powerful tools in aromatherapies to induce recall, reinforce powerful emotion, or simply for the calming quality of their fragrances.

When they affect us in these profound ways, it is because we resonate with them.

On a visual level we are attracted to them by their stunning beauty, and we display them in our homes and businesses to "brighten things up." It is a long held tradition to bring roses to a romantic partner to show love and as a traditional courting strategy. We also send flowers to commemorate special occasions like birth, marriage, and death.

These visual and olfactory cues combine to attract pollinators like hummingbirds and bees who stimulate other synergistic interactions to spread seeds and enhance survival strategies aided by other species, and this is only what we see in the frequency spectrums that we normally perceive. Who knows what might be happening in the spectrums that other species can perceive that we cannot?

Similarly, in the vegetable kingdom, as shown in the complex interaction of the olfactory and taste receptors, these characteristics combine to help us identify what is healthy to ingest and aid us.

On vision quests and in visionary states, particularly those encountered on strict shamanic plant diets centered around Ayahuasca, subjects imbibe any number and combination of specialized plants to learn from the spirits of the plants by passing through physical, mental, and spiritual ordeals to prove that they are worthy of the gift of knowledge that the plant spirits have to give them.

Each plant and animal has its own spirit, essence, or energy which can be characterized as its own unique personality in the same way that Ayahuasca is universally referred to as "The Mother." It is also the reason why North American Indians don't refer to animals as the bear or the coyote. They say Bear or Coyote as they consider them all to be manifesting the entire essence of their spirit in a grouped manner, similar to bees, ants, and other cooperative colonies, and each one has its own unique energy signature that is treated with equal respect like the well-known personality of Coyote the Trickster.

When you ingest a particular plant or plants, especially in a specific environment while on a strict cleansing diet, their effects are strongly felt and enhanced, even more so the longer the diet is continued. During that time participants are subject to the energy field of that plant or plants, which all interact with each other and our bodies and minds in different ways.

The combination of the diet, environment, the plants, and other elements affect our brain wave activities and physiologies in a myriad of mysterious ways on physical, psychological, and spiritual levels that we can scarcely comprehend, much less study in any traditional objective scientific way. The best way to study them is from the outside in through the subjective perceptions they produce in experiences that often transpire on deep non-rational levels of resonance and synchronization with the plants as transmitters in the position of orchestra conductors, and the subjects as receivers, communicating in ways that defy logic and articulation.

An Ayahuasca ceremony is characterized as an individual healing experience in a group setting with much anecdotal evidence supporting sporadic telepathic experiences among participants. The healing circle is referred to as a container to safely hold the energy it generates and attracts so that the participants can be protected while in the vulnerable state necessary for healing and release.

If a group of people all sat around with their own radios tuned to the same station listening to the same song, all of the radios would be playing the same music as one, in sync. If all of the participants in the healing circle tuned in and resonated at the same frequency or frequencies, there is no reason why they could not have similar shared telepathic events.

This learned ability to subject themselves to these diverse natural energies to surrender and discover what mysteries they may reveal puts shamans in the humbling position of being subject to their plant teachers as the conductor transmitters, in spite of the physical or psychological discomfort they may have in the process.

In the Amazon, Ayahuasca is often referred to as *la purga,* because it purges the body and rigorously clears it out through every orifice, including vomiting, defecation, profuse sweating, tear ducts, and the lungs, as well as energetically at deep psychological and spiritual levels. Often it is used in conjunction with other intensely purgative plants along with the restricted diet referred to as the *dieta* consisting of boiled rice, oatmeal, or quinoa, baked or boiled unripe bananas, and chicken or fish once a day or less combined with a pitcher a day of a helper plant or plants. There is no salt, no soap, shampoo, toothpaste, scents or additions of any kind, and no sex. In addition to the deep inner purification, other plants are used for daily plant baths to physically and energetically wash away the toxins and energies that are released in the process.

All of these restrictions enhance the cleansing process over the extended time of the *dieta,* which dates back to prehistory. As time passes subjects lose their telltale human scents, particularly pheromones, hormones, and other secretions that jaguars and other animals with a highly refined sense of smell can detect, making the hunters virtually invisible to them and their primary sensory modalities. The constant refined ingestions and plant baths result in the subject not only smelling like the jungle, but they *become* the jungle, giving them a distinct hunting advantage.

These intense practices fine tune the subject's own senses bringing great clarity and a highly refined perception of their immediate reality, further adding to their hunting advantage. Aside from the intensified state of awareness that comes from this extreme purification which historically prepared them for hunts, battles, or other challenges, their conscious awareness of their field of perception expands at inner and outer levels making them psychologically and mentally clear for the task at hand.

These hard won physical and perceptual enhancements bring deep inner shifts from the resonance that comes from being sympathetic to and in sync with the plant spirits and the energies they manifest. One of the intriguing results of this immersion into what shamans characterize as the spirit world is the agreed upon encompassing energetic field that opens up the ability to not only commune with the plants and the distinctive energies of their unique spirits and personalities, but also to the distinctive energies and personalities of the animal kingdom.

This phenomenon is reinforced by widespread reports from Ayahuasca drinkers of directly experiencing and communing with or being "possessed" by specific animal energies that are common to the Ayahuasca experience, regardless of whether they are in the jungle or in a major North American city.

Among the numerous totems claimed by participants, or which they say actually choose them, the most common are condors, jaguars, and snakes which have a deeper meaning going all the way back to prehistoric cultures. Those experiencing these energies or spirits often roar and growl through no volition of their own like jaguars, flap their legs like wings, or feel their bodies swaying seemingly of its own accord to distinctive serpentine movements. Other animals and insects like hummingbirds, butterflies, dragonflies, dolphins, and other aquatic totems can play big parts as well.

This phenomenon with its roars, "wing flapping" and other indicators that an animal spirit is present or "possessing" its host forms the core of shape shifting mythologies where shamans reportedly have the ability to change their form into that of their familiar animal totems.

In this way shamans share the distinct, intimate resonant vibration of these entities in the same energetic way they do with the morphogenetic field of plants, becoming one with them by sharing the same frequency. In the same way they have to surrender to the plant teachers to discover what mysteries may be revealed, shamans must humble themselves to their animal familiars as the conductor transmitters of their experience.

In the lore of the jungle, by sharing in the vibratory field of the animal's spirit energy, in a state of surrender on that entity's terms, that animal learns from the human by seeing things through human eyes while the human learns other modes of perception from seeing things through the animal's eyes.

In this manner, by mutual agreement and the understanding that comes from sympathetically sharing the same energetic field and perception of specific animals, combined with the clarity that comes from the physical purifications that makes the hunters virtually invisible to the primary sensory modalities of their prey, the shaman hunts the animals in spirit first, then follows through what has happened in the spirit world in the physical world in the actual physical hunt, aided by his cultivated invisibility.

This learned ability of shamans to tune in to and commune with these diverse plant and animal energies that we all share the world with allows them to experience them in a direct, definitive, and subjective manner that helps them understand other perspectives. This cultivates empathy which is the ability to understand and share the feelings of another, and opens them up in ways that only direct experience can, while aiding them in developing the critical skills necessary for what is defined as soul retrieval.

In all of my experiences outside of consensual time and space the normal rules of perception no longer apply. Colors with hues that defy description bombard me and unfold in multicolored geometric progressions that could be microcosmic quantum expressions, or unfolding galaxies. Within these realms I have lived as an insect devoured by still bigger insects, which have in turn been devoured by lizards and snakes with long ethereal stomachs that have passed me into nonrational dimensions that both amaze and terrify.

Outside of my physical body the frogs, birds, insects, jaguars, and other creatures of the Peruvian Amazon fill the night air with their calls, cries, twitters, and buzzes. For me there is no difference between the infinity expressing itself outside of me and the infinity that I soar through inside of me. It is all one. Outside of time and space a noise from deep in the jungle sounds as if it is right beside me, startling me. Sometimes I feel myself fully present and aware in two places at the same time, often in different times and dimensions.

After experiencing the consciousness of predator and prey in the lower worlds, I have flown first as a Condor, then as a Hummingbird into sublime and exquisite high frequency realities exploding with neon luminescent pastel manifestations that defy description. While my spirit soars, my body quivers and my insides teeter on the verge of both vomiting and shitting.

After being subjected to infinite varieties of altered states of consciousness, non-ordinary realities, and any number of earthly and otherworldly intelligences, including those in my day to day waking life, I now experience my furthest reaching travels and adventures with my constant, trusted, deeply loved hummingbird spirit, Picaflor, and it is to this defining spirit of my existence that I dedicate this book to.

PICAFLOR FLIRTS FIRST

The second time I went to the Amazon to do a shamanic plant *dieta*, I worked with an older *vegetalista* named Guillermo who was a mentor to the *ayahuasqueros* I was working with. Guillermo sang powerful, intense, focused songs that made my whole body twitch and tremble. In one ceremony I was sitting cross-legged when my legs started going up and down on their own accord, whoop, whoop, whoop, first like the flapping of butterfly wings, then like massive bird wings that I believed to be the wings of a condor. My whole body spasmed and jerked with bird-like movements and in my visions I flew through a high mountain range both visualizing and feeling first like a butterfly, then a condor at the center of creation flying through valleys and soaring upwards looking at the stars. It was all so spectacularly beautiful that any attempt at expression falls woefully short of articulating the experience.

Guillermo sang to me and blew tobacco and cinnamon into the top of my head and all over my body. He blew on my stomach, sucked at it, and spit out phlegm or energy, then he talked and sang into my stomach and blew into my hands a number of times, into my feet, and on my legs, then he continued blowing tobacco into my stomach and the rest of me so I closed my eyes and went with the whole experience.

I started coming down as he made the rounds of the rest of the circle. Someone was yelling and growling like an animal, which he said later was his jaguar totem. The lady beside me twitched, spasmed and banged the floor while I did my best to stay quiet. My legs continued flapping like massive wings for the longest time as if carrying me along through ineffable portals and dimensions of what I thought of as shamanic soul travel.

The next morning when Guillermo brought me my day's dose of a second teacher plant I wondered about his icaros and how my flapping legs became the flight of a butterfly that transformed into a condor, but I didn't say anything. He put his arms out wide, gave me a big smile and flapped his arms like wings as if he was flying which confirmed for me that he and his songs had a part in taking me there. When I flapped my own arms and asked him about my flight his smile widened and he said, "El Condor!"

The following year when I was headed South again I commented to my girlfriend that morning about the hummingbirds I saw at the fountain in our yard before carpooling to L.A. with a friend. We stopped at the Getty museum on the way. While walking I mentioned the hummingbirds I saw that morning and seconds after I spoke a hummingbird darted in and flew around in front of us for a moment. Stunned by the synchronicity my friend and I looked at each other with wide-eyed amazement.

I had brief flashes of higher frequency energy over the next few years that in retrospect felt like flirtations coupled with intensely beautiful visions that were almost too much to bear until one night, deep in my visions my usual Condor "wing" legs started flapping faster than ever before and I became overwhelmed in an experience that felt outside of time and infinite. My body twitched and my head bobbed while I flew through neon pastel colored bliss into sublime, exquisite high frequency luminescent realities while my body quivered at a high speed that I did not think possible.

I felt honored by the attention from what I thought of as a playful conflict that hinted at jealousy between the slower energy of Condor and this higher energy and at one point it felt like I held both energies within me. Fascinated, I worked with that dance of energies until the higher vibration took over and pulled me up out of the bigger, slower moving Condor without any urge to vomit or defecate. I felt pure joy and believe my body's energies had adjusted to embracing these high frequencies and raised up enough in vibration to handle and embrace them fully.

I felt exceptionally clear and open the following morning when we met to discuss our experiences and subsequent integration, and the full impact of what happened hit me when a friend said, "I knew we were going to have a powerful session because just before I came inside a hummingbird flew up to me and hovered in front of my face."

The moment he said that I realized I had embodied the spirit of hummingbird and in that same moment of revelation my girlfriend beside me exclaimed, "You were the hummingbird!"

I felt a profound shift inside of me in that flash of insight and a mystical inner rebirth, and understood that I had reached a new level of awareness and would never be the same from that moment going forward.

I felt deeply honored that Picaflor had chosen me and a little embarrassed that it had taken me so long to embrace this loving realization, but I don't think my body and its energies were ready for its full embrace before this, which would explain its short flirting visits in life and in the jungle in my visions leading up to this momentous blessing. I had been loyal to Condor and believe that all my time with her cleared me out and raised my vibration to prepare me for fully merging and accepting Picaflor. Now after all these years Picaflor is my primary totem while my deep connection, respect, and gratitude with Condor remains and she still joins me from time to time in my visionary journeys.

Those who know me in shamanic circles call me Picaflor.

SPIRIT AMBASSADORS

Someone once told me that hummingbirds are the nerve endings of God, and ever since that dynamic image has been indelibly tattooed into my brain. Everything about them manifests high frequency energy as well as every aspect of their physiology, including their speed, metabolism, vision, hearing, and possibly other forms of higher perception outside the range of humans and other animals. Even the myriad of legendary sparkling jewel colors they reflect speak of high frequency light as do the visions I have when Picaflor flies with me through neon pastel colored bliss into sublime, exquisite high frequency luminescent realities while my body quivers at a high speed that does not seem possible.

Hummingbirds have close to a thousand feathers on their bodies. Their sparkling iridescence comes from light reflected from these feathers which have black melanin granules that function as color pigments. In most birds, the melanin granules are randomly scattered in the feather structure, but in hummingbirds, the melanin granules are hollow and flattened like pancakes stacked neatly in seven to fifteen rows.

When light enters a feather some colors are absorbed by the black melanin granules and others like red in the case of ruby-throated hummingbirds are scattered back out to an observer's eye.

The angle of the light that passes through the stacks of melanin granules changes the distance that photons of light move and the changing angles alter the red color that is reflected back, so when a hummingbird turns its head, the gorget seems to change from one shade of red to another in a shimmering fashion which is how iridescence works in hummingbirds.

If the sunlight hits the gorget at a particularly flat angle, the color that is reflected is dark, almost black. When the hummer turns to a more favorable angle with the sun, the red pops out.

The shape and size of the melanin granules determine the color that is reflected iridescently. The Blue-throated Hummingbird found in Arizona and points south has granules that produce a breath-taking blue iridescence.

The sunlight that shines on a hummingbird's feathers reflects this transcendent beauty and the life giving sun that is the source of this is composed of hydrogen and helium, the top two elements of the periodic table. These primary elementals are manifestations of energy that interact to create the light and heat that the sun radiates unconditionally.

In another aspect or manifestation of light, or better still *lightness,* what is lighter than air? A hummingbird's speed, agility, and mastery of movement through the air pushes the limits of the known boundaries of physics. Their wings move back and forth horizontally, drawing a narrow, elegant figure eight in the air with each full stroke which is continuous like a Mobius strip, the symbol of infinity. Seventy-five percent of its lift is produced during the down stroke. The figure-eight pattern is similar to a swimmer doing figure eights in the water to remain afloat. Imagine making a figure-8 with your arms 80 times within one second. This isn't humanly possible, but it is for hummingbirds.

It is well known that the primary elemental spirits in ancient cultures are earth, air, fire, and water, and traditional Chinese medicine adds metal to this list. If you examine these primary elementals in terms of their density, or rate of vibration defined as frequency, air and fire are the two highest frequency elementals and are pure manifestations of energy with no discernable mass.

Elemental spirits are universally revered in shamanistic cultures and one of the most respected among them is the Wind, an invisible force that can be harnessed as well as storm out of control, leaving death and destruction in its wake. The fact that it is capable of delivering so much power, yet has no substance, defines it as not only a direct link to spirit, but spirit itself; formless, invisible, energy. If we remain still, we have no perception of air when it is at rest and we perceive and characterize it as *nothing* as it registers nothing with any of our senses, making it a perceptual enigma.

Air is the most pervasive presence we know. It surrounds and caresses us both inside and out, moving across our skin, between our fingers, around our arms and thighs, sliding along the roof of our mouths and down our throats to fill our lungs to feed our blood, which keeps us alive.

We cannot speak, act, or think without the participation of this fluid element and we exist in its depths the same way fish live immersed in the ocean. We can feel it moving against us and often taste, smell, and hear it as it swirls in our ears or moves through whispering leaves, or when it changes the shape, moves shifting clouds, or sends ripples along the surface of a lake. The fluttering feathers of a hummingbird, a spiraling leaf as it falls, and the slow drift of a seed through space indicate the presence of the air, yet we can never see the air itself.

This unseen enigma holds the mystery that enables life to live and unites our breathing bodies to the world around us and to the interior life of all that we perceive in the open field of this living presence, and what the plants breathe out, we and the animals breathe in, and what we breathe out plants breathe in. In shamanic thought, the air is the soul of the visible landscape that constitutes the secret realm where all beings draw their nourishment from. As the mystery of our living present, it is that most intimate absence felt as nothing where the present presences, providing a key to the forgotten presence of the earth.

Nothing is more common to the diverse indigenous cultures of the earth than a recognition of the air, wind, and breath as aspects of a singularly sacred power. By virtue of its pervasive presence, it's invisibility and its influence on all manner of visible phenomena, the air for oral people is the archetype of all that is ineffable, yet undeniably real and efficacious. Its obvious ties to speech in that spoken words are structured breath, and broken phrases take their communicative power from this invisible medium that moves between us lends the air a deep association with linguistic meaning and thought. Its ineffability is similar to the ineffability of awareness itself. Many indigenous people construe awareness or mind not as a power that resides inside their heads, but as a quality that they themselves are inside of along with the other animals, plants, mountains, and clouds.

In his book, *Holy Wind in Navajo Philosophy*, James Mikhail McNely asserts that the Navajo term *nilch'i*, meaning Wind, Air, or Atmosphere, suffuses all of nature and is that which grants life, movement, speech, and awareness to all beings. Additionally, the Holy Wind serves as the means of communication *between* all beings and elements of the animate world, making *nilch'i* central to the Navajo worldview.

For the Navajo *nilch'I* is a single unified phenomenon of the Wind in its totality comprised of many diverse aspects of partial Winds, each with their own name in the Navajo language. One of these - *nilch'i hwii'siziinii*, or the "Wind within one", refers to that part of the overall Wind that circulates within each individual. The Wind within one is not autonomous, but a continual process of interchange with the various winds that surround us, and is a part of the Holy Wind itself.

When referring to the multiple Winds like Dawn Man, Dawn Woman, Sky Blue Woman, Twilight Man, Dark Wind, Wind's Child, Revolving Wind, Glassy Wind, Rolling Darkness Wind, and others, the Navajo are not speaking of abstract or ideal entities. These entities are not palpable phenomena like gusts, breezes, whirlwinds, eddies, storm fronts, cross currents, gales, whiffs, blasts, and breaths that they perceive in the fluid medium that surrounds and flows through their bodies.

The Navajo conviction that all of these subsidiary Winds are internal expressions of a single, inexhaustible mystery comes from the observation that the multiple vortices made by their own breathing, heat rising in waves, or the branches of trees as they sway in the surging air. All these currents and eddies swirling around and inside them are not entirely autonomous forces, but momentary articulations within the vast and fathomless body of Air itself.

For the Navajo the air in its capacity to provide awareness, thought, and speech has properties that European alphabetic civilization traditionally ascribe to an interior, individual human mind or psyche, yet by attributing these powers to the Air, and insisting that the "Winds within us" are continuous with the Wind at large, with the invisible medium that we are immersed in, Navajo elders suggest that what we call "mind" is not ours, and therefore not a human possession. Mind as Wind is a property of the encompassing world that humans and all other beings participate in. One's individual awareness and the sense of a personal self or psyche is simply that part of the enveloping Air that circulates within, through, and around one's body, so one's own intelligence is assumed from the start to be entirely participant with the swirling psyche of the land.

Our English term psyche and its modern offspring psychology, psychiatry, and psychotherapy are derived from the ancient Greek word *psychê*, which denoted not just the soul or the mind, but also breath or a gust of wind. The Greek noun was itself derived from the verb *psychein*, which meant to breathe or blow while another ancient Greek word for air, wind, and breath, *pneuma*, gives us pneumatic and pneumonia, while signifying that vital principle which in English we call "spirit."

The word spirit itself, despite all its incorporeal and non-sensuous connotations is directly related to the bodily term respiration through their common root in the Latin word *spiritus*, which meant breath and wind, as well as being the root of the words inspire and inspiration. Similarly the Latin word for soul, *anima* gives us animal, animation, animism, and unanimous, which is being of one mind or one soul. It also signified air and breath. These were not separate meanings. *Anima*, like *psychê*, originally named an elemental phenomenon that comprised both what we now call the air and what we now term the soul. The more specific Latin word *animus* signified that which thinks in us was derived from the same airy route, *anima*, itself derived from the older Greek term *anemos*, meaning wind.

We find an identical association of the mind with the wind and the breath in numerous ancient languages. Even an objective, scientifically respectable word as atmosphere displays its ancestral ties to the Sanskrit word *atman*, which signified soul as well as the air and the breath. A great many terms that refer to the air as a passive and insensate medium are derived from words that once identified the air with life and awareness. Words that now seem to designate a strictly immaterial mind or spirit are derived from terms that once named the breath as the very substance of that mystery.

For ancient cultures the air was a singular sacred presence. As the experiential source of both psyche and spirit, it appears as if the air was once felt to be the very matter of awareness, the subtle body of the mind, and that awareness, far from being experienced as a quality that distinguishes humans from the rest of nature was felt as that which invisibly joined human beings to the other animals, plants, forests, and mountains as the unseen common medium of their existence.

Like so many other ancient tribal languages, Hebrew has a single word for both "spirit" and "wind", the word *ruach*, the spiritual wind, which is central to early Hebraic religiosity. Its primordiality and close association with the divine is evident in the first sentence of the Hebrew Bible:

When God began to create heaven and earth - the earth being unformed and void, with darkness over the surface of the deep and a wind (*ruach*) in God's sweeping over the water...

At the beginning of Hebrew creation, God is present as a wind moving over the waters, and breath, as we learn in the next section of Genesis, is the most intimate and elemental bond linking humans to the divine. It is that which flows most directly from God and man, for after God forms an earthling (*adam*) from the dust of the earth (*adamah*), he blows into the earthling's nostrils the breath of life, and the human awakens.

Although *ruach* refers to the breath, the Hebrew term used here is *neshamah* which denotes both the breath and the soul. While *ruach* generally refers to the wind or spirit at large, *neshamah* signifies the more personal, individual aspect of the wind, the wind or breath of a particular body like the "Wind within one" of the Navajo. In this sense *neshamah* also signifies conscious awareness.

The ancient Hebrews were among the first communities to make sustained use of phonetic writing and the first bearers of an alphabet. Unlike other Semitic peoples they did not restrict their use of the alphabet to economic and political record keeping. They used it to record ancestral stories, traditions, and laws, possibly making them the first nation to so thoroughly shift their sensory participation away from the forms of surrounding nature to a purely phonetic set of signs that made them experience profound epistemological independence from the natural environment made possible by this potent new technology. To actively participate with the visible forms of nature came to be considered idolatry by the ancient Hebrews. For them it was not the land, but the written letters that now carried their ancestral wisdom.

Although the Hebrews renounced all animistic engagement with the visible forms of the natural world, whether with the moon, sun, or animals like the bull, which were sacred to other peoples of the Middle East, they retained a participatory relationship with the invisible medium of that world with the wind and the breath.

The power of this relationship can be inferred from the structure of the Hebrew writing system, the *aleph-beth*. This ancient alphabet, in contrast to its European derivatives had no letters for what we call "vowels." The twenty-two letters of the Hebrew *aleph-beth* were all consonants, so to read a text written in traditional Hebrew one had to infer the appropriate vowel sounds from the consonantal context and add them when sounding out the written syllables.

One of the primary reasons for the absence of written vowels in the traditional *aleph-beth* has to do with the nature of vowel sounds themselves. While consonants are shapes made by the lips, teeth, tongue, palate, or throat that momentarily obstruct the flow of breath and gives form to our words and phrases, vowels are those sounds made by unimpeded breath itself.

Vowels are nothing other than sounded breath, and the breath for the ancient Semites was the very mystery of life and awareness inseparable from the invisible *ruach*, holy wind, or spirit. Breath was the vital substance blown into Adam's nostrils by God himself who granted life and consciousness to humankind. It is possible that the Hebrew scribes refrained from creating distinct letters for the vowel sounds in order to avoid making a visible representation of the invisible. To fashion a visible representation of the vowels of the sounded breath would have concretized the ineffable and make a visible likeness of the divine. It would have created a visible representation of a mystery whose essence was to be invisible and unknowable - the sacred breath, the holy wind, so it wasn't done.

The absence of written vowels marks a profound difference between the ancient Semitic *aleph-beth* and the subsequent European alphabets. Unlike texts written with the Greek or the Roman alphabets, a Hebrew text could not be experienced as a substitute for the sensuous, corporal world. The Hebrew letters and texts were not sufficient in and of themselves. In order to be read, they had to be added to, enspirited by the reader's breath. The invisible air, the same mystery that animates the visible terrain was also needed to animate the visible letters to make them come alive and speak. The letters themselves remained dependent upon the elemental corporeal life world that they were activated by, the very breath of that world, and could not be cut off from that world without losing their power.

In this manner the absence of written vowels ensured that Hebrew language and tradition remained open to the power of what exceeded the strictly human community, ensuring that the Hebraic sensibility would remain rooted, however tenuously, in the animate earth. While the Hebrew Bible became a kind of portable Homeland for the Jewish people, it could never take the place of the breathing land itself, upon which the text manifestly depends, hence the persistent themes of exile and longed-for return that reverberate through Jewish history down to the present day.

The absence of written vowels in ancient Hebrew entailed that the reader of a traditional Hebrew text actively choose the appropriate breath sounds or vowels and different vowels frequently varied the meaning of the written consonants.

The apparent precision and efficiency of the new alphabet was obtained at a high price. For by using visible characters to represent the sounded breath, Greek scribes effectively desacralized the breath and the air. By providing a visible representation of that which was by its very nature invisible, they nullified the mysteriousness of the enveloping atmosphere negating the uncanniness of this element that was both here and yet not here, present to the skin, yet absent the eyes, immanence and transcendence all at once.

Plato and Socrates were able to co-opt the term *psychê*, which had been fully associated with the breath and the air, employing it to indicate something not just invisible but intangible. The Platonic *psychê* was not at all a part of the sensuous world, but of another non-sensuous dimension. The *psyche* was no longer an invisible yet tangible power continually participant by virtue of the breath with the enveloping atmosphere, but a thoroughly abstract phenomenon now enclosed within the physical body as in a prison.

Hummingbirds are not only masters of the air, but they are primarily made of air. Their tiny bodies are crammed with no fewer than nine air sacs, in addition to their two huge lungs and enormous heart. Their delicate bones are exceptionally porous, their legs are thinner than toothpicks, and their feet are as flimsy as embroidery thread.

Native American Hummingbird Art

MYTH AND MAGIC

Hummingbird symbolism varies from culture to culture. They have different meanings for individuals and universal meanings to people around the world, among them joy, healing, beauty, sweetness, good luck, variety, angels, spirits, messengers, flirtatiousness, and agility.

There are over 330 species of hummingbirds found only in the Western Hemisphere, with almost half of them living in an equatorial belt between 10 degrees north and south of the equator. They come in a vibrant array of jewel-like colors and are associated with the colorful flowers they get nectar from and pollinate, making them symbols of vibrancy, variety, and flirtatiousness because they fly quickly from blossom to blossom, enjoying the nectar of a variety of flowers. Unlike ravens and eagles hummingbirds do not mate for life. They have co-evolved with flowers in a harmonious partnership where they get nectar from the flowers and in return pollinate more flowers, which extends their life force which is why hummingbirds are considered symbols of health, healing, and vitality.

Hummingbirds are believed by many to be messengers from spirit guides, something that I have found to be true in my own experience, particularly when they accompany me in visionary journeys. According to Native American and other shamanic traditions, you don't choose your spirit animals; they choose you. I can't begin to express how loved, honored, flattered, and blessed I feel to have been chosen by Picaflor and how right and fitting it feels so deep within my heart.

Tens of millions of years ago, hummingbirds lived in the geographical region that is now Europe, but in more recent history they are native only to the Americas with a range from southern Alaska all the way to Tierra del Fuego at the southern tip of South America. They can also be found in the Caribbean Islands. Because of this range, human cultures of the Americas and tropics are rich with hummingbird legends and folklore.

Native Americans revere nature and animals, and their culture has many vivid stories about the power of animal spirits and the profound role they play in human life. While Native American cultures are diverse, among them the hummingbird is viewed as a helpful spirit guide and messenger by many tribes, as well a healer and a source of good luck.

In legends, Hummingbird is often portrayed as a healer or as a spirit being who helps people in need, and sometimes plays the important mythological role of fire-bringer. Seeing a hummingbird is a sign of good luck in many Native American tribes, especially Northwest Coast tribes. In ancient Mexico, hummingbirds were considered sacred and associated with royalty and warriors; the Aztec god Huitzilopochtli, patron god of the Aztec capital Tenochtitlan had the hummingbird as his divine animal and was sometimes depicted in hummingbird form in traditional Aztec art. In some Mexican tribes today, hummingbirds are believed to be messengers from the afterworld or manifestations of a dead person's spirit.

Hummingbirds are also used as clan animals in some Native American cultures. Tribes with Hummingbird Clans include the

Pueblo tribes of New Mexico. Hummingbird is an important clan crest in some Northwest Coast tribes, and can sometimes be found carved on totem poles.

What follows are myths and legends from many of the indigenous people of the Americas that date back to prehistory.

THE HUMMINGBIRD STORY OF THE HURON AND IROQUOIS

Long ago, before humans roamed the earth, it was always daylight and all the animals lived in peace. No one remembers why exactly they started fighting with each other, but The Great Spirit was not pleased, and tried to talk to the animals to stop them from fighting. The animals were so loud and causing such a ruckus, they didn't even hear the Great Spirit giving them good advice.

Very angry, the Great Spirit threw a blanket across the sky and thus it was now dark all the time. The animals went into despair and put away their petty fighting that no one knew the cause of anyway.

The animals held counsel and were trying to come up with a solution to the problem. First the bear with mighty paws said I am strongest and I will rip down this blanket. Bear climbed to the highest mountain and jumped as high as possible and could only swipe at the blanket, making a huge streak across the sky. The bear fell down the mountain, defeated.

The animals kept arguing who would go next. The little hummingbird offered, but everyone ignored such a little animal. The vulture decided to try and flew up with a beautiful head of feathers that poked a hole in the blanket. Vulture came down screaming as the sun had burned all those beautiful feathers right off.

The animals continued to argue and the hummingbird snuck away, gathering as much might as a hummingbird can, flying fast all the way up to the sky. Hummingbird poked and poked that tiny beak through the blanket and light shined through, on and on and on, finally falling back in exhaustion. The animals gathered around the little hummingbird and were saddened that they bickered and bickered and here the smallest yet mightiest of them all just went and tried to do their best.

The Great Spirit was pleased and told the animals that light would return to the world but, the blanket would go back on every day at the end to remind them to not fight. This created night and day. The bear's swipe can be seen in the Milky Way, the vulture's hole is now the moon, and the hummingbird left the greatest gift of all; stars!

A NAVAHO HUMMINGBIRD
STAR CREATION MYTH

A long time ago there lived four animals on earth. Each of the animals lived peacefully and in turn Great Spirit rewarded them with eternal sunlight, but one day something new came into the world - fighting and jealousy. Atop a large tree grew a very big and delicious looking fruit and upon seeing it, Bear began climbing the trunk of the tree to retrieve the beautiful thing. Coyote, also seeing the fruit, yearned for it and jumped onto Bear's back, bringing both animals crashing to the ground. As they began to fight about who the fruit belonged to, Eagle flew between the branches of the tree and plucked the fruit from where it sat. As he landed, thinking himself very clever to have won the fruit from the other two, he began to laugh, dropping the fruit from his beak and onto the soft forest floor; so ensued the greatest fight that animals have ever been a part of.

So disruptive was this fight that its noise carried high into the sky and into Great Spirit's ears. Angered by chaos below Great Spirit threw his cape upon the world, bringing, for the first time, an immense darkness. A silence immediately fell on the earth below, as the animals stopped their fighting in search of the source of this new thing. Unable to see anything, unable to understand what had happened, but knowing that they had angered Great Spirit, they began fighting once again, accusing one another of their own faults. Finally it was decided that one animal must go to Great Spirit and ask for forgiveness.

"I'll do it," said Coyote. "I will swim the widest river and climb the highest mountain and I will yell to Great Spirit, asking for forgiveness."

And so Coyote set off across the planes of the world, swam the widest river and climbed the highest mountain until he reached its top. And there he stood and yelled to Great Spirit.

"Please forgive us. We have learned our lesson!" But there was no reply. For hours Coyote stood and yelled until his voice became hoarse and tired, emitting nothing but a howl, and so he returned to the others.

Bear laughed at his defeated friend, "You thought you could yell to Great Spirit?" And so Bear set off. Upon reaching the top of the Great Mountain, Bear took in a deep breath, filling his chest with air, and yelled as loud as he could, "Please forgive us, Great Spirit!" But there was no reply. After hours of failed attempts, Bear returned to his friends, grumbling with frustration.

Upon his return, Eagle laughed with joy at his friend's failure. "Now you will see what a true animal can do." But no sooner had he begun his climb towards the heavens did he become very tired. A rush of death enveloped his body and he returned to earth, choosing his life over the happiness of his friends.

The mightiest of Great Spirit's animals had fallen and with their realization of this, a severe cold began to inhabit the animals. Then, out of the darkness, they heard a tiny voice, "I can do it." And from the darkness emerged Hummingbird.

The animals could not believe their eyes - never had they seen this tiny, pathetic creature before. They immediately burst into a fit of laughter. "You will never reach the Great Spirit my little friend," said Bear. "I am the great warrior of this land and I was unable to make my pleas heard. Stay here where it is safe, for if you set off you will know only defeat."

Hummingbird ignored the jests of his fellow animals and began his flight. He flapped his wings and as he reached the Great Spirit's cape his long beak punctured the fabric, but his body gave way and he fell back down to earth. Looking up into the sky the animals saw little Hummingbird's work, for a tiny pinprick of light was now showing through the cape. After a short break, Hummingbird tried again, and again a pinprick appeared in the cape just before he fell back to earth. Finally, after thousands of attempts, Hummingbird made one last try - flapping his wings faster than ever as he made his way into the darkness. One last hole appeared in the sky, but Hummingbird did not. After several minutes, Great Spirit appeared before the animals with tiny Hummingbird's body clutched in his hands.

Hummingbird's heart so impressed the Great Spirit that in honor of the fallen animal the Great Spirit covers the earth with his cape, un-mended, for half of the day, every day.

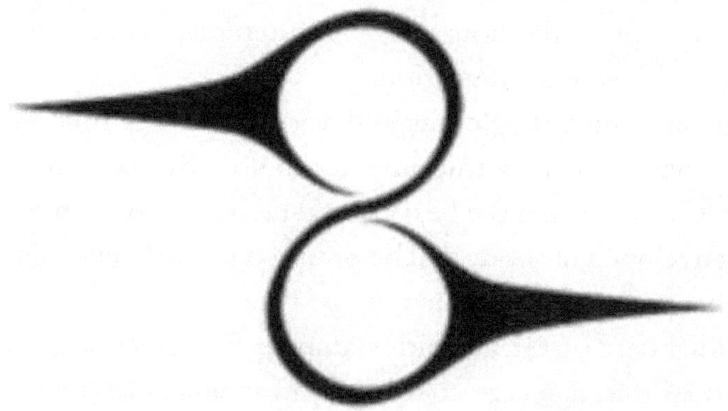

Navajo and Cherokee Symbol For Hummingbird (dah yi'itihi)

Navajo Hummingbird Kachina Doll

HOW HUMMINGBIRD GOT FIRE
A RUMSIEN OHLONE STORY

The Ohlone are Native people from the Central California Coast, from the San Francisco Bay through Monterey and down to the Salinas Valley.

Once, a very long time ago, something happened to the world. The ocean rose up higher and higher, covering the land little by little until nearly everything was covered over with water. Eagle, Hummingbird, Crow, Raven, and Hawk were together on a mountaintop, looking out at the world and seeing how it had been destroyed.

After many days, Eagle had an idea and used her magic and help from Hawk to dry up the waters. By this time, the five friends were hungry. There was food to be found, but they needed fire to cook with. Eagle knew the one place where fire could be found, and sent her little nephew Hummingbird to get fire from the Badger People underground, but the Badger People refused to share their fire and sent Hummingbird away.

When Hummingbird returned, Eagle was very angry and sent him back. This time, the Badger People saw Hummingbird coming and said, "Cover the fire! Cover the fire!" They hid their fire by covering it over with deer skin, but the deer skin had a hole in it where an arrow had gone through, and Hummingbird reached in with his long, narrow beak. He took out a hot ember and carried it away, but before he could put it safely under his wing, it flamed, turning his throat brilliant red. That's why Hummingbird has a red throat, and that's how there came to be fire in the world again.

Pacific Northwest Native American Hummingbird Art

Pacific Northwest Native American Hummingbird Art

HOPI AND ZUNI LEGENDS

Hopi and Zuni legends tell of hummingbirds intervening on behalf of humans, convincing the gods to bring rain. Because of this, people from these tribes often paint hummingbirds on water jars. The Hopi kachina for Hummingbird depicts him with green moccasins and a green mask. He has an aqua body, and he is yellow on top of the head. He is crowned with a ruff made of Douglas fir.

There is a story told in the Hopi and Zuni tradition of a famine that plagued the land. Because food and water were scarce, a mother and father were forced to leave their young son and daughter behind as they went searching for supplies to sustain their family.

To pass the time while waiting, the boy carved a hummingbird out of a piece of wood. When his sister threw the toy into the air the small bird came to life and played with the children. Seeing that they were hungry, the hummingbird began to worry about them, so she went searching and returned every day with an ear of corn to sustain them. Realizing they would need more to eat, the hummingbird flew to the center of the Earth to beg the God of Fertility to replenish the land. Impressed by the beautiful and sincere little bird, the God of Fertility delivered rain, which fed the soil and helped the crops to grow again.

Zuni Hummingbird Carving

Hopi Hummingbird Kachina

Antler Hummingbird Zuni Fetish

APACHE FOLKLORE

In Apache folklore, there is a story of a young warrior named Wind Dancer who is deaf. Despite his deafness, Wind Dancer creates beautiful, wordless songs. Indeed, the sound is so exquisite that the songs bring fair weather and healing to the land.

One day, Wind Dancer falls in love with a beautiful woman named Bright Rain. They meet when Bright Rain is being attacked by a **wolf** and Wind Dancer steps in to rescue her, but early in their romance, Wind Dancer is killed tragically in battle. When he dies, winter falls on the land.

Grieving, Bright Rain walks into an open field and when she does, the spirit of Wind Dancer visits her in the form of a hummingbird. The tiny bird sings Wind Dancer's sweet, wordless songs in Bright Rain's ears and she is comforted in her grief. When Bright Rain feels a sense of peace, fair weather returns to the land.

Apache Legend Artwork

Apache Artwork

A CHEROKEE LEGEND

HUMMINGBIRD BRINGS BACK TOBACCO

Long ago, when all people and animals spoke the same tongue, there was only one tobacco plant in all the world. From far and wide, they came for their tobacco. All was well until the greedy Dagul'ku geese stole the plant and flew far to the south with it where they guarded it with all their might. Before long all of the people and the animals began to have great sufferings because they had no tobacco. One old woman who had long suffered became so thin and weak that it was held by all that she would die soon, and could only be saved by tobacco.

This old woman was loved by all of the people and the animals and this disturbed them greatly so it was decided that they should hold a council, which they did, and make a plan on how to return the tobacco that had been taken from them. They decided to send the animals to retrieve it. One by one, the animals all tried to get the plant, but each time they were seen by the Dagul'ku geese and killed before they could reach the plant. From the largest to the smallest, the four footed animals failed.

Now, the Mole spoke up and said that he would go. Everyone thought this to be a good idea, as he could tunnel under the ground to the plant and steal it away. So, off he went and as he approached the plant, his track was seen by the Dagul'ku who waited at the plant for him to come out. When at last, out he did come, he suffered the same outcome.

Much disparity was heard in the council after that. No one could think of any way to get the tobacco plant away from the greedy Dagul'ku. No other animals wanted to go. The Hummingbird had been listening to all of the plans and came up with one of his own. At last he told the council that he could retrieve the plant. They looked at him and said how could this be, you are so small? How could you get to the plant past the Dagul'ku? He told them that he could do it and that if they wanted they could test him. So out in the middle of the meadow they showed him a plant that all could see. They said to him, "Go, sit on that plant, but do not let us see you getting there."

No sooner than the words had been spoken did all of the people see the Hummingbird sitting atop the plant in the meadow, and right before their eyes he disappeared again and reappeared in the council circle. Not one person saw him go or return. Stunned at the feat, it was decided to give him a chance. He wasted no time and off he dashed straight to the plant, right under the noses of the Dagul'ku. Right up to the plant without them even suspecting him being there. Quick as a wink, he used his long beak to cut off the top of the plant that had a few leaves and seeds, then off he dashed straight back to the council circle. By this time the old woman was thought to have died, but smoke was blown into her nostrils and with a cry of "Tsa'lu", she opened her eyes and regained her strength.

Cherokee Hummingbird Art

Cherokee Hummingbird Art

THE RACE BETWEEN THE CRANE
AND THE HUMMINGBIRD

The Hummingbird and the Crane were both in love with a pretty woman. She preferred the Hummingbird, who was as handsome as the Crane was awkward, but the Crane was so persistent that in order to get rid of him she finally told him he must challenge the other to a race and she would marry the winner. The Hummingbird was so swift--almost like a flash of lightning--and the Crane so slow and heavy, that she felt sure the Hummingbird would win. She did not know the Crane could fly all night.

They agreed to start from her house and fly around the circle of the world to the beginning, and the one who came in first would marry the woman. At the word the Hummingbird darted off like an arrow and was out of sight in a moment, leaving his rival to follow heavily behind. He flew all day, and when evening came and he stopped to roost for the night he was far ahead, but the Crane flew steadily all night long, passing the Hummingbird soon after midnight and going on until he came to a creek and stopped to rest at daylight.

The Hummingbird woke up in the morning and flew on again, thinking how easily he would win the race, until he reached the creek and found the Crane spearing tadpoles with his long beak for breakfast. He was very much surprised and wondered how this could have happened, but he flew swiftly by and soon left the Crane out of sight again.

The Crane finished his breakfast and started on, and when evening came he kept on as before. This time it was hardly midnight when he passed the Hummingbird asleep on a limb, and in the morning he had finished his breakfast before the other came up. The next day he gained a little more, and on the fourth day he was spearing tadpoles for dinner when the Hummingbird passed him.

On the fifth and sixth days it was late in the afternoon before the Hummingbird came up, and on the morning of the seventh day the Crane was a whole night's travel ahead. He took his time at breakfast and then fixed himself up as nicely as he could at the creek and came in at the starting place where the woman lived, early in the morning. When the Hummingbird arrived in the afternoon he found he had lost the race, but the woman declared she would never have such an ugly fellow as the Crane for a husband, so she stayed single.

A PUEBLO INDIAN TOBACCO LEGEND

Most Native American tribes are based around a clan system, which is a community organization rooted in maternal family lines. Each clan is associated with a specific animal. For example, the Creek, Chippewa, Algonquian, Navajo, and Pueblo Nations all have bear clans. The Pueblo Nation is one tribe that has a hummingbird clan. Thus, these special birds have important meaning in Pueblo stories.

According to one Pueblo legend, caterpillars are the guardians of tobacco plants, but it's the hummingbird who gathers smoke from the caterpillar and brings it to the medicine men who use it to purify the Earth. In return, the hummingbird brings gifts from the shamans to the Great Mother who exists beneath the soil.

Zia Pueblo Indian Pottery Hummingbird Jar

Kewa Pueblo Indian Pottery Hummingbird Jar

THE HUMMINGBIRD AND THE RAINBOW

There was a demon who made a bet with the Sun and lost. When he lost the bet, he went blind. Filled with anger at going blind, the demon spewed hot lava, which set the world on fire.

The hummingbird, who was a simple gray bird at the time, took it upon himself to fly around and gather rain clouds from all four directions and put out the fire. When his work was done he flew away, but as he did there was a rainbow in the sky which he flew through. The rainbow blessed the hummingbird by painting him with its bright colors. To this day, members of the Pueblo Nation have hummingbird dances and use hummingbird feathers in rituals to bring rain and Pueblo shamans use hummingbirds as couriers to send gifts to the Great Mother who lives beneath the earth.

PIMA AND TARASCAN WATER LEGENDS

In a Pima legend a hummingbird acted like Noah's dove, bringing back a flower as proof the great flood was subsiding.

There is also a legend from Mexico about a Tarascan Indian woman who was taught how to weave beautiful baskets by a grateful hummingbird to whom she had given sugar water during a drought. These baskets are now used in Day of the Dead Festivals.

Tarascan Hummingbird Art

AZTEC LEGENDS

For the ancient Aztecs, the God of Sun and War was called Huitzilopochtli which translates to Hummingbird Wizard and combines the Aztec word for hummingbird and sorcerer who spits fire. Ancient Mexicans called them huitzitzil and ourbiri, rays of the Sun and tresses of the day star.

For the Aztecs, Huitzilopochtli was a very important patron god. In fact, hummingbirds were so sacred in Aztec culture that only shamans and tribal leaders were allowed to wear hummingbird feathers. In addition, Aztec warriors believed that if they died in battle they would be reincarnated as hummingbirds, who are furious fighters who have returned to life with swords as beaks, continuing their battles forever in the sky. The Aztecs admired hummingbirds so deeply that they adorned statues of Montezuma with their feathers.

Early Spanish visitors to the new world, seeing hummingbirds for the first time called them resurrection birds and believed that anything that glittered so brightly had to have been made new each day.

Quetzalcoatl **with Hummingbird (Creation)**

AN AZTEC CREATION MYTH

In the beginning was only Tepeu and Gucumatz, the Feathered Serpent who also had the name Quetzalcoatl. These two sat together and thought, and whatever they thought came into being. They thought earth, and there it was. They thought mountains, and so there were. They thought trees, and sky, and animals etc., and each came into being. None of these things could praise them so they formed more advanced beings of clay, but these beings fell apart when they got wet, so they made beings out of wood, but they proved unsatisfactory and caused trouble on the earth. The gods sent a great flood to wipe out these beings so that they could start over. With the help of Mountain Lion, Coyote, Parrot, and Crow they fashioned four new beings. These four beings performed well and are the ancestors of the Quich.

In the beginning there was only darkness. Suddenly a small bearded man, the One Who Lives Above, appeared rubbing his eyes as if just awakened. The man, the Creator, rubbed his hands together and there appeared a little girl, Girl-Without-Parents. The creator rubbed his face with his hands and there stood the Sun-God. Again Creator rubbed his sweaty brow and from his hands dropped Small-boy. Now there were four gods, then he created Tarantula, Big Dipper, Wind, Lightning-Maker and Lightning-Rumbler.

All four gods shook hands so that their sweat mixed together. Then Creator rubbed his palms together from which fell a small round, brown ball. They took turns kicking it and with each kick the ball grew larger. Creator told Wind to go inside the ball and blow it up. Then Tarantula spun a black cord which he attached to the ball and went to the east pulling as hard as he could.

He repeated this exercise with a blue cord to the south, a yellow cord to the west and a white cord to the north. When he was done the brown ball had become the earth. The Creator again rubbed his hands and there appeared Hummingbird. "Fly all over this earth," said Creator to Hummingbird, "and tell us what you see." When he returned Hummingbird reported that there was water on the west side, but the earth rolled and bounced, so Creator made four giant posts one each black, blue, yellow and white and had Wind place them at the four cardinal points of the earth. The earth was now still. The creation of the people, animals, birds, trees, and everything else took place after.

Aztec Hummingbird Art

TWO MAYA LEGENDS

One legend of the Mayan hummingbird tells us that the gods created all animals and gave each one a specific job to do on earth. When they thought they had completed the job, they discovered that they had forgotten to provide a messenger to transport their thoughts and desires from one place to another.

When the gods realized their mistake, they also discovered they had run out of mud and maize. Without the materials to create their new messenger, they had to get creative. Fortunately, as gods, they were creators of the impossible.

For this new creation, the Mayan gods decided to create their messenger with something special. They took a jade stone and began to carve in it an arrow. This arrow was designed to represent a journey. After a few days, the stone arrow was ready. When they blew on the stone to get rid of the dust caused by their carving, they blew so hard that it flew into the sky. As the arrow flew, it turned into a beautiful multicolored hummingbird which they called x ts'unu'um.

The gods new messenger was fragile and so light it could fly around a person without them realizing it. They could approach the most delicate flowers without causing them any harm. It was this lightness of being that made the hummingbird the perfect carrier for man's thoughts and desires. These thoughts and desires were carried without being noticed.

According to this Mayan legend, the hummingbird became so important that people tried to capture the bird and exploit their power. The gods were so angered by this disrespectful behavior they condemned to death any man who dared cage one of these precious creatures. Additionally, they endowed the hummingbird with great speed to evade their wannabe-captors. This is one of the mystical beliefs as to why it is so difficult to capture the hummingbird, due to their protection by the gods.

These birds are also thought to be able to carry messages from beyond this life. In this way, they are manifestations of a spirit of the deceased. The hummingbird is also considered a mythological healing animal that helps people in need by changing their luck.

The Mayan legend of the hummingbird also claims that this precious, tiny and stealthy bird has the important job of taking people's thoughts and intentions from one place to another. When someone saw a hummingbird, it was believed that this was the manifestation of positive thoughts from someone who cares about them. It could even represent love.

Another Mayan legend says the first two hummingbirds were created from the small feather scraps left over from the construction of other birds. The god who made the hummers was so pleased he had an elaborate wedding ceremony for them told in the following myth.

Tzunuum, the hummingbird, was created by the Great Spirit as a tiny, delicate bird with extraordinary flying ability who could fly fast. When the hummingbird flew above the Creator, her wings made a humming sound: "dzu-nu-ume, dzu-nu-ume', so the Mayas called her Dzunuume, or the hummer. She was the only bird in the kingdom who could fly backwards and who could hover in one spot for several seconds. The hummingbird was very plain. Her feathers had no bright colors, yet she didn't mind. Tzunuum took pride in her flying skill and was happy with her life despite her looks.

When it came time to be married, Tzunuum found that she had neither a wedding gown nor a necklace. She was so disappointed and sad that some of her best friends decided to create a wedding dress and jewelry as a surprise. Ya, the vermilion-crowned flycatcher wore a gay crimson ring of feathers around his throat in those days. He decided to use it as his gift. So he tucked a few red plumes in his crown and gave the rest to the hummingbird for her necklace.

Uchilchil, the bluebird, generously donated several blue feathers for her gown. The vain motmot, not to be outdone, offered more turquoise blue and emerald green. The cardinal, likewise, gave some red ones, then Yuyum, the oriole, who was an excellent tailor as well as an engineer, sewed up all the plumage into an exquisite wedding gown for the little hummingbird. Ah-leum, the spider, crept up with a fragile web woven of shiny gossamer threads for her veil. She helped Mrs. Yuyum weave intricate designs into the dress. Canac, the honeybee, heard about the wedding and told all his friends who knew and liked the hummingbird. They brought much honey and nectar for the reception and hundreds of blossoms that were Tzunuum's favorites.

Then the Azar tree dropped a carpet of petals over the ground where the ceremony would take place. She offered to let Tzunuum and her groom spend their honeymoon in her branches. Pakal, the orange tree, put out sweet-smelling blossoms, as did Nicte, the plumeria vine. Haaz, the banana bush, Op, the custard Appletree, and Pichi and Put, the guava and papaya bushes made certain that their fruits were ripe so the wedding guests would find delicious refreshments. Finally, a large band of butterflies in all colors arrived to dance and flutter gaily around the hummingbird's wedding site.

When the wedding day arrived, Tzunuum was so surprised, happy, and grateful that she could barely twitter her vows. The Great Spirit so admired her humble, honest soul that he sent word down with his messenger, Cozumel the swallow, that the hummingbird could wear her wedding gown for the rest of her life, and to this day, she has.

Maya Hummingbird Art

White Wind Mayan Rainbow Hummingbirds

**Corazón del viento Colibrí Acrílico /Lienzo
Heart of the Wind, Hummingbird**

THE HUMMINGBIRD KING
A GUATEMALAN FOLKTALE

The Hummingbird King is a Guatemalan folktale that tells the tragic story of Kukul, the son of a chief to whom a beautiful bird appears at birth, leaving behind a unique red feather which becomes his protective charm. As the young man grew he fought in battles and wars. Not once was he struck with an arrow. Due to his bravery and leadership, the townspeople made him their chief, but his uncle Chirumá was jealous and desired the position for himself.

His uncle snuck up on him while he slept and stole his red feather, leaving him unprotected, then he shot Kukul with an arrow the following day while he was hunting in the forest, but magically Kukul transformed into a beautiful green hummingbird with scarlet chest, the quetzal that became a symbol for the Mayan people, and remains so for those of Central American origin today.

TAINO LEGENDS

The Taino Nation are an Indigenous People from the Caribbean, who revered the hummingbird as a symbol of fertility for all life. According to Taino legends, the hummingbird started out as a fly, but Father Sun, whom they call Agueybaba, transformed him into a tiny bird.

For the Taino, the hummingbird is a symbol of rebirth, and even though they are small, they have the fierce heart of an eagle. The Taino People call their warriors the *"Colibri"* Warriors, which translates to Hummingbird Warriors.

In other Caribbean folklore, hummingbirds are believed to be spirit messengers, including guides and departed loved ones.

Taino Caribbean Indian Hummingbird Art

A Hummingbird on a Caribbean Airlines jet

The two birds on the shield of the Coat of Arms of the Republic of Trinidad and Tobago are hummingbirds. Trinidad is sometimes referred to as the "Land of the Hummingbird" because more than sixteen different species of hummingbird have been recorded on the island. "Land of the Hummingbird" is also believed to have been the Native Amerindian name for Trinidad.

A QUECHUA PARABLE

My painter friends from the Peruvian Amazon tell me that Tonkiri, their name for hummingbird, was the great teacher who taught all the animals, geniuses, and spirits of the forest the art of color, since in the beginning of the creation of the Amazonian world everything was gray.

The following hummingbird parable originated with the Quechuan people of South America, and has become a talisman for environmentalists and activists worldwide committed to making meaningful change.

This popular story of the hummingbird is about a huge forest being consumed by a fire. All the animals in the forest come out and they are transfixed as they watch the forest burning and they feel very overwhelmed, and very powerless, except this little hummingbird. It says, "I'm going to do something about the fire!" So it flies to the nearest stream and takes a drop of water. It puts it on the fire, and goes up and down, up and down, up and down, as fast as it can.

In the meantime all the other animals, much bigger animals like the elephant with a big trunk that could bring much more water, are standing there helpless, saying to the hummingbird, "What do you think you can do? You are too little. This fire is too big. Your wings are too little and your beak is so small that you can only bring a small drop of water at a time."

As they continue to discourage it, the hummingbird turns to them without wasting any time and tells them, "I am doing the best I can."

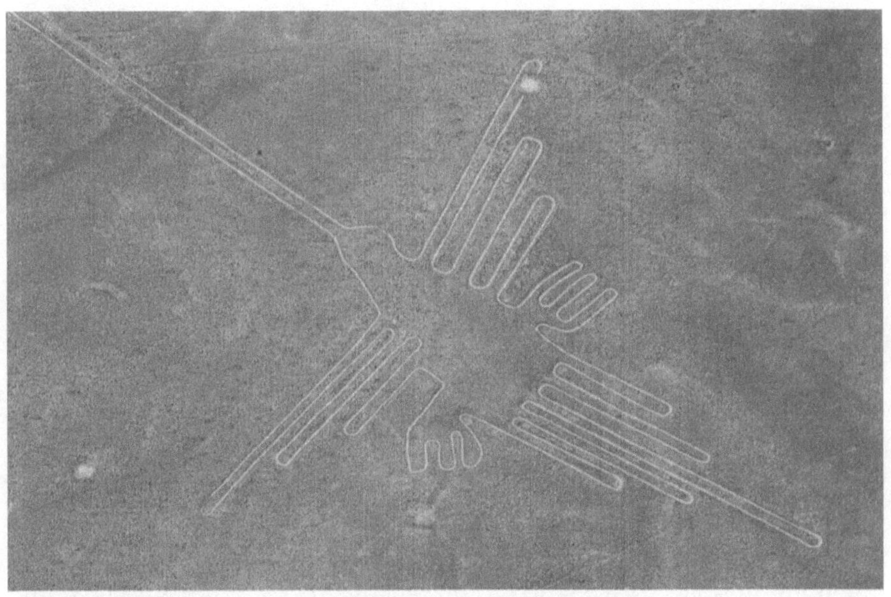

Nazca Lines Peru

INCA LEGENDS

The Inca believed that the hummingbird was a messenger from heaven. The condor, which has the position of "king of the skies" conceded its status as the primary spiritual messenger of the "upper world" to the hummingbird. The hummingbird is seen as the key to the next stage of development of human consciousness.

Incan legend tells that there was a contest between the condor and the hummingbird to see who would be king. The condor said he could fly to the edge of the sky and the hummingbird said he could fly beyond the edge to the center of heaven. When it came time for the contest, the hummingbird didn't show up. The condor took off and flew to the edge of the sky at which point the hummingbird emerged from the condor's feathers and flew beyond to the center of the upper world where he met Wiraqocha the metaphysical god of the Andes.

This Inca myth about Picaflor is the first one I ever heard and for me the one that holds the most relevance, particularly in light of my own deep experience of first connecting with Condor before transitioning to Picaflor. This magical experience first came in flashes of higher frequency energy that felt like flirtations coupled with intensely beautiful visions that were almost too much to bear until one night, deep in my visions my usual Condor "wing" legs started flapping faster than ever before and I became overwhelmed in an experience that felt infinite and outside of time. My body twitched and my head bobbed while I flew through neon pastel colored bliss into sublime, exquisite high frequency luminescent realities while my body quivered at a high speed that I did not think possible.

Inca Cosmic Hummingbird Ceramic Plate

Inca Ceramic Earthtone Vessel: Hummingbird Feast

Incan Hummingbird

Inca Hummingbird Pin (Tupu)

FACTS AND FIGURES

The bodies of hummingbirds contain so many opposites that they're very existence is a paradox. They are the lightest birds in the sky, and the fastest for their size. These tiny, diaphanous birds undertake perilous, long distance migrations like that of the Rufus hummingbird which flies from Mexican wintering grounds to nesting areas in Alaska.

Hummingbirds can execute acrobatics that no other bird can approach. Alone among the world's ten thousand avian species, only hummingbirds can hover in mid-air. For centuries nobody knew how they did it and their physics defying feats were considered pure magic.

Birds are physiologically very different from humans and our fellow mammals who are all fluid filled creatures. Early Greek physicians believed all medicine could be based on an understanding of these fluids they called humors, but in order to fly birds cannot afford to be loaded down with heavy fluids. Birds, *especially* hummingbirds are made of air.

Unlike our thick marrow filled bones, most bird's bones are hollow. Even their skulls are scaffolded with passageways for air. Their feather shafts are hollow, and the feathers themselves, like strips of velcro, are interlocking barbules for catching air. Their bodies are filled with air sacs, which originate in and function in part as extensions of the lungs. No fewer than nine of these filmy bladders fill the tiny body of a hummingbird, one pair in the chest cavity, another under each shoulder blade, another pair in the abdomen, one under each wing, and one along the neck. Hummingbirds are the lightest birds in the sky. Of their roughly 240 species, all confined to the Western hemisphere, the largest, an Andean giant is only 8 inches long while the smallest, the bee hummingbird of Cuba, is just over 2 inches long and weighs a single gram.

When a hummingbird swoops in you know it thanks to the unmistakable bold humming sound it produces from wings that flap so fast they're little more than a blur.

Researchers at Stanford designed a system comprised of microphones, cameras, and pressure sensors sensitive enough to detect air pressure changes created by Picaflor flapping wings that generated a "3D acoustic model" that revealed the secrets of hummingbird hums.

The hum is generated from the pressure difference between the topside and underside of the wings, which changes both in magnitude and orientation as the wings flap back and forth. Larger birds produce a "whooshing" sound when they fly created by the powerful but slow flapping that keeps them aloft. Hummingbirds hum because their wings move so fast that there is pressure on both the up and downstroke of the wing, allowing the bird to hover while blurring the sound of each flap into a hum.

"This is the reason why birds and insects make different sounds," David Lentink, co-author of the study, said. "Mosquitoes whine, bees buzz, hummingbirds hum, and larger birds 'whoosh'. Most birds are relatively quiet because they generate most of the lift only once during the wingbeat at the downstroke. Hummingbirds and insects are noisier because they do so twice per wingbeat."

Delicacy is what gives Picaflores their unrivaled powers of flight. Alone among birds, they can hover, fly backward, even fly upside down, and for such small birds their speed is astonishing. In his courtship display to impress a female, a male Allen's hummingbird, can dive out of the sky reaching 61 miles per hour, plunging from 50 feet at a rate of more than 60 feet per second. Pulling out of his plunge he experiences more than nine times the force of gravity. Adjusted for body length, the Allen's is the fastest bird in the world. Diving at 385 body lengths per second this hummer beats peregrine falcon dives at 200 body lengths per second, and even bests the space shuttle as it screams down through the atmosphere at 207 body lengths per second. At more than 60 times a second, hummingbird wings beat at a rate that makes them a blur to humans and for centuries people deemed hummingbird flight pure magic.

Until the invention of the stroboscope, scientists could not understand how hummingbirds hover. With a flash duration of one hundred thousandths of a second, the stroboscope finally revealed the motion of wings that had been too fast for other cameras to capture. Picaflores are less flesh than fairies and they are not much more than bubbles fringed with iridescent feathers.

AIR WRAPPED IN LIGHT

Touching hummingbirds damages their feathers as they mostly consist of air surrounding a humongous heart. Feathers are made of keratin, the same as human fingernails, horse's hooves, and rhino horns, but the keratin in feathers, due to a difference in molecular structure is tougher.

A typical bird's feathers outweigh its skeleton and feathers define a bird. By trapping and moving air, feathers protect the bird from cold and wet, and enable it to fly, but each feather is largely air, with a stiff central shaft that is light, hollow, and attaches beneath the skin to a muscle. Like each scale on a reptile, each feather on a bird can be raised or lowered as needed. The shaft divides the feather into two broad veins on each side consisting of parallel branches called barbs. At right angles to the barbs are interlocking branchlets called barbules. Hundreds of pairs of tiny barbules on each barb fit together like tiny strips of velcro that give feathers their web like quality. The barbules have small hook like processes called barbicels and on the barbicels of some feathers are tinier branchlets, microscopic hamuli, that allow even more air to be trapped in the feather. When Picaflores preen, running their beaks through their feathers, they are rezipping the Velcro of these minute connections.

Caught in the feathers, air gives birds their warmth and flight. In hummingbirds, air even gives them their color. Their jewel-like radiance, emerald, ruby, amethyst, does not come from pigment like most bird's feathers, but from air. Except for the flight feathers on wings and tail, the top third of hummingbird feathers lack barbicels and hamuli. These have elliptical structures called platelets which are different from the clotting cells in our blood of the same name, filled with microscopic air bubbles. The shape and thickness of the platelet in the amount of air determine the color scene. These air bubbles diffract light into colors that reflect back in flashes of iridescence.

Hummingbirds pack more feathers per inch than any other bird with the possible exception of the penguin, but the hummingbird's feathers are far more festive. The bright feathers on their sparkly throat patches called gorgets scatter light like soap bubbles, creating some of nature's most spectacular colors.

The color on the throat, or gorget, and head of a male is particularly spectacular. The platelets on these feathers are like flat mirrors, and light reflects in only one direction which is why the gorget of a male ruby throat dazzles in sunlight, but may look black in shade. Hummingbirds know this. By carefully adjusting his position and relation to the observer and the sun, a male purposely flashes his colors to intimidate a rival or attract a mate.

The simplest gliding flight of birds demands precise balance and judgment and exploits passive lift to counteract the pull of gravity. Vultures, hawks, and eagles ride thermal currents on rising columns of warm air. Albatrosses and petrels exploit different layers of wind speed above waves. Birds can glide for hours, expending very little energy.

Flapping flight, the way most birds fly, is more demanding, achieved by flexing wings at joints in wrist, elbow, and shoulder powered by extraordinarily strong breast muscles. The wings move forward in a downward arc, propelling the bird forward and up similar to the oarsman's power stroke or the action of a swimmer doing the butterfly. Movement flows into the upward stroke, then a recovery stroke to start the process anew.

Hovering is unique to hummingbirds. No other bird hovers. Kites, storm petrels, kestrels, and kingfishers appear to do so, but only hummingbirds can sustain this method of flying for more than a few moments. Instead of flapping their wings up and down, the wings move forward and backward in a figure 8. During the forward and back strokes, the wings make two turns of nearly 180 degrees. The upstroke as well as down stroke require enormous strength, and every stroke is a power stroke. Like insects and helicopters, hummingbirds can fly backward by slanting the angle of the wings, and can fly upside down by spraying the tail to lead the body into a backward somersault. Hovering becomes so natural to a hummingbird that a mother who wants to turn in her nest does it by lifting straight up into the air, twirling, then coming back down. A hummer can stay suspended in the air for up to an hour.

In most birds, 15 to 25 percent of the body is given over to flying muscles. In a hummingbird's body, flight muscles account for 35 percent and an enormous heart constitutes up to 2.5 percent of its body weight, the largest per body weight of all vertebrates. At rest, the hummingbird pumps blood at a rate 15 times as fast as that of a resting ostrich, and the blood is exceptionally rich in oxygen carrying hemoglobin.

Hummingbirds are some of the smallest birds in the world and are the fastest fliers relative to their body length, and they are the only true hoverers. To enable this special form of flight, hummingbirds have evolved several distinct adaptations from a specialized wing shape to breast muscles that take up about 30 percent of their entire body weight while most bird's breasts weigh in at around 15-18 percent.

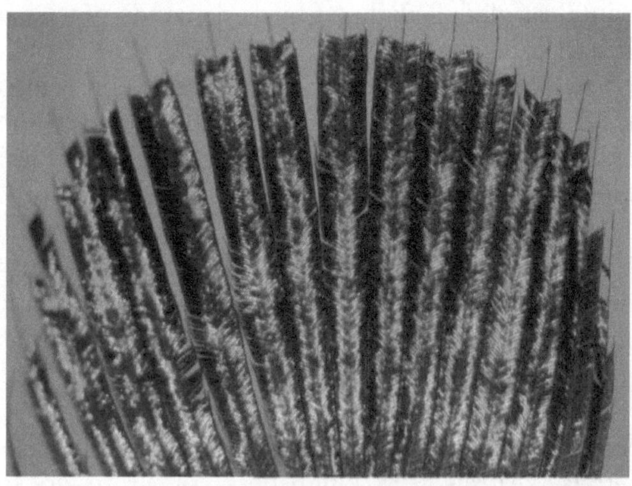

A close-up of a Ruby-throated Hummingbird feather.

Magnified feather close up

Increased magnification feather close up

Extreme magnified feather close up

Wine-Throated Hummingbird Gorget

Gorget Close up

Magnified gorget close up

BRAINS AND BRAWN

Hummingbirds have brains unlike any other bird or four-limbed vertebrates which lets them manage multidirectional flight.

To understand these differences, it's important to consider how and why we move. Every creature on Earth is either predator or prey, and many are both depending on where they fall in the food chain. For almost all animals, this means moving forward, either toward food or away from becoming food. Humans are a rare exception among vertebrates in that we have the capability to move with relative efficiency in many directions.

Despite this difference, our first instinct when faced with danger is to turn around and run away from the threat, not waltz sideways to safety. This instinct can be traced to the way our brains perceive motion. The visual centers in the brains of all four-limbed creatures have the strongest response to motion on the back-to-front axis which is evident when chasing something or being chased.

Hummingbirds spend a lot of time hovering, which means they have more to consider than just the front back-to-front axis. When hovering, a gust of wind might push them from the side, or a predator might strike from below, so they have to be able to move not just forward to feed on a dangling flower, but in all directions.

Picaflor brains don't put the same emphasis on back to front movement that ours do. Researchers have discovered that in an area of the brain called the lentiformis mesencephali, the part that responds to visual stimuli, hummingbirds don't have a strong back to front preference like all other animals they tested. Picaflores seemed to have no preference and responded to motion equally in all directions.

Researchers also found that hummingbird brains are tuned to be more responsive to fast movements than slow ones. This was a surprise because scientists had assumed that their brains would be tuned for a low-speed hover, but being optimized for high speed makes sense. An Anna's Hummingbird can move at 385 body lengths per second during mating flights, which is the highest known length-specific velocity attained by any vertebrate. In comparison, an F15 Eagle fighter jet has a top speed of Mach 2.5, which translates to around 45 body lengths per second. At that speed the ability to make near-instantaneous course corrections is the difference between life and death, or mating and not mating, which is the same thing in evolutionary terms.

The fact that hummingbird brains perceive the world differently than other vertebrates is a fascinating ornithological find. The discovery of an animal brain that can move efficiently in three dimensions could have great value for artificial intelligence in flying drones or computerized autopilot systems for helicopters. Commercial applications aside, it's interesting to consider that the tiniest bird brains very well could be the most complicated.

A hummingbird brain making up 4.2 percent of its weight is proportionally the largest of any bird's. By comparison, our brains are 2 percent of our body weight. Inside that big little brain is an encyclopedia of important information. Picaflores can remember every flower they've ever visited, including on migration routes. They can figure out how long to wait between visits so the flowers have time to generate more nectar. They can even recognize humans, and know which ones can be counted on to refill empty feeders.

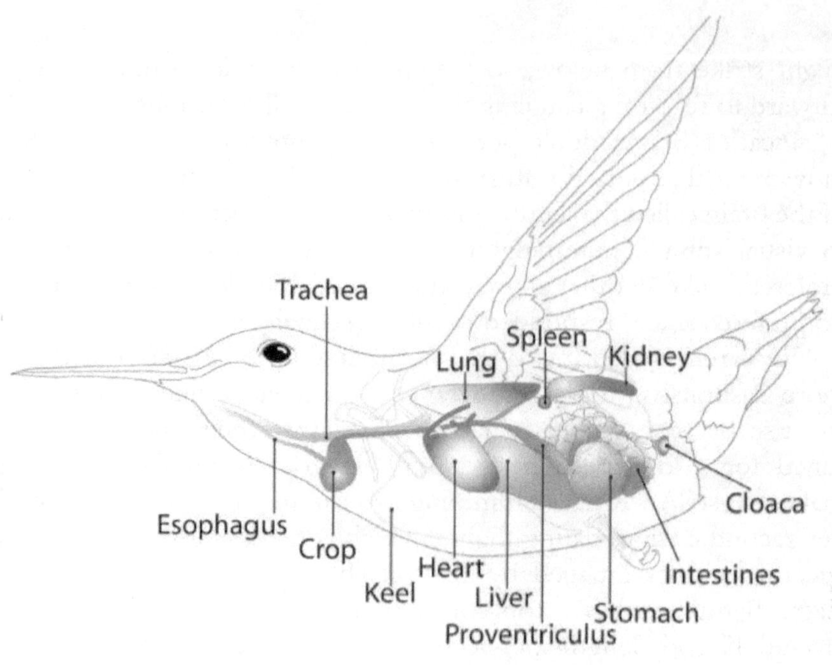
Trachea
Spleen
Lung
Kidney
Esophagus
Crop
Cloaca
Keel
Heart
Liver
Intestines
Stomach
Proventriculus

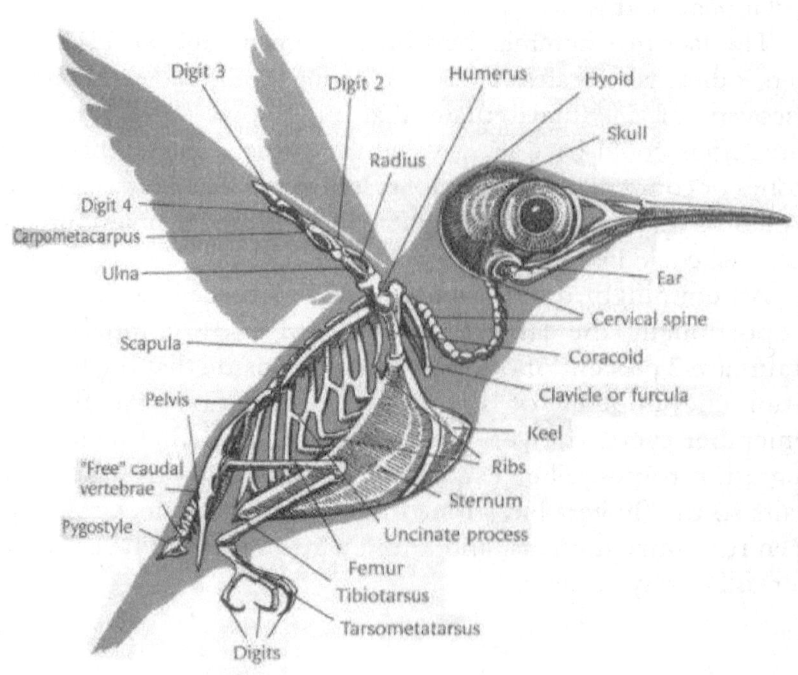
Digit 3
Digit 2
Humerus
Hyoid
Radius
Skull
Digit 4
Carpometacarpus
Ulna
Ear
Cervical spine
Scapula
Coracoid
Clavicle or furcula
Pelvis
Keel
"Free" caudal
vertebrae
Ribs
Sternum
Pygostyle
Uncinate process
Femur
Tibiotarsus
Tarsometatarsus
Digits

BY THE NUMBERS

Rufous hummingbirds make long migrations that take them up and down the length of North America each year and they have the ability to keep track of particularly juicy flowers depending on where they appear, whether first, second, or even fourth in a line-up of blooms. This understanding of numerical order is a complex skill that may help hummingbirds remember the easiest routes between nectar-rich flowers.

Many animals can count, and some understand how things fit together in a sequence. Rats, guppies, and monkeys trained in a lab can all use sequences to find food, but this doesn't tell us how wild animals might use this ability in a natural setting.

Susan Healy, a biologist at the University of St. Andrews, and colleagues studied rufous hummingbirds which weigh less than a nickel and are just 8 centimeters long, with well-defined feeding territories and excellent memories of what's on their turf.

These birds use efficient routes to head from one nectar-rich flower to another like a shopper carefully planning the best route through a grocery store. Healy's team wanted to find out if they simply move from one visual target to the next one in sight, or did they learn a sequence, knowing which items follow the current one?

The researchers set up feeders and once they saw a bird consistently eating from a certain feeder and defending his territory from other birds, they trapped and marked him for identification, then trained nine marked hummingbirds to feed from an artificial flower consisting of a yellow foam disc on a wooden stake with a syrup containing tube in the center.

To see whether the animals had a sense of numerical order, the researchers lined up 10 identical artificial flowers. When they put syrup in the first flower to see where the hummingbirds went to feed the birds went almost uniformly to the first flower, sometimes giving the others a quick check to see whether they might also hold a tasty treat.

The researchers then rearranged the flowers after each visit, mixing them up, even moving the entire line so that the position of the flowers couldn't give the birds information about which flower had the syrup. Even then, the birds chose the first flower in the line, suggesting they had a concept of "first." When the team repeated the entire experiment and baited the third flower, the birds usually zoomed straight toward it, suggesting that they knew the third flower in line had the treat, regardless of where the line actually was.

THE SOUND OF LIGHT IN SONGS

Consisting of chirps, squeaks, whistles, and buzzes, hummingbird songs originate from at least seven specialized nuclei in the forebrain. A genetic expression study showed that these nuclei enable vocal learning through imitation, a rare trait known to occur in only two other groups of birds; parrots and songbirds, and a few groups of mammals, among them humans, whales, dolphins, and bats. Within the past 66 million years, only hummingbirds, parrots, and songbirds out of 23 bird orders may have independently evolved seven similar forebrain structures for singing and vocal learning, indicating that evolution of these structures is under strong epigenetic constraints possibly derived from a common ancestor.

The blue throated hummingbird's song differs from typical songbirds in its wide frequency range, extending from 1.8 kilohertz to approximately 30 kilohertz. It also produces ultrasonic vocalizations which do not function in communication. As blue throated hummingbirds often alternate singing with catching small flying insects, it is possible the ultrasonic clicks produced during singing disrupt insect flight patterns, making insects more vulnerable to predation.

In the Andes, Ecuadorian Hillstar hummingbirds chirp songs of seduction only other birds of its kind can hear. As the male sings, he inflates his throat creating a wave-like motion that causes iridescent throat feathers to glisten purple. His ballad is sung at around 13.4 kilohertz which is considered ultrasonic for birds, which generally can't hear above 9 or 10 kilohertz.

Among birds, only some owls have been shown to hear ultrasonic sounds which they use to locate prey, but not to communicate.

Researchers went into the Ecuadorian Andes to reach high grasslands to locate the Hillstars' breeding grounds and record the males singing, then played back their romantic ballads to test the reactions of other birds.

Other Hillstars craned their necks and turned toward the speaker as it played the high-pitched chirps and one of them flew over the speaker to inspect it. In the lab, the scientists verified that the part of the brain typically engaged in auditory communication had been activated.

Mammals generally hear a wider range of pitches than birds. Humans can hear pitches up to about 20 kilohertz, but lose sensitivity to high-pitch sounds with age.

The researchers believe the birds may have evolved to sing at high pitches so their love songs wouldn't compete with background noises in their environment like mountain winds, streams, and the songs of other birds.

Claudio Mello, a behavioral neuroscientist from Oregon Health and Science University, visited the Brazilian forest in 2015 armed with a battery of sensitive microphones and recording equipment usually used to record the high-frequency calls of bats. In prior years, when conducting research in the region, he had heard a high-pitched sound on the edge of hearing that he concluded, amidst the clamor of life in the Brazilian forest, probably came from a hummingbird rather than an insect or a tree frog.

He discovered that black jacobins were uttering high-frequency calls at a level beyond belief. Most birds hear in the range of two to three kilohertz. Humans typically hear noises in the range of one to four kilohertz. The black jacobins were repeatedly broadcasting calls above ten kilohertz and up to fourteen kilohertz, far beyond the regular range known for any bird, let alone a small hummingbird.

This was widely reported as black jacobins uttering the highest-pitched calls of any bird in the world, but that accolade belongs to the blue-throated hummingbird. Its song included elements that exceeded twenty kilohertz, but there was no evidence that the blue-throated hummingbirds could hear anything beyond seven kilohertz, while the black jacobins high-pitched calls vocalized almost exclusively above ten kilohertz.

This data indicates that they can hear their own kind, but why have they evolved a call that was so far out of the range of hearing for the remainder of the bird kingdom?

Mello hypothesized that this might be an adaptation to their biodiverse surroundings. Black Jacobins are found in a geographic range that includes over forty other hummingbird species and subspecies, not to mention myriad other voices in the forest. They may have evolved to have their own private frequency that eliminated much of the background noise. Some Australian geckos have exceptionally high-frequency calls and hearing, while the hole-in-the-head frog, a species of frog endemic to Borneo, communicates entirely in the ultrasonic range up to thirty-eight kilohertz. If reptiles and amphibians found evolutionary advantage in communicating in high frequencies, it follows that some birds might have done so too.

A SONG AND DANCE FOR THE LADIES

While Black Jacobins have evolved to shout out loud to catch one another's attention, another hummingbird species evolved to let their feathers do the talking. Anna's hummingbird, the species that, in the US, is expanding its range north on the coattails of climate change, has a courtship display flight that at first glance is merely visually arresting. The male ascends about thirty meters into the air above his prospective partner, tucks his wings into his body, and then plummets in a fast dive toward her, pulling up at the last possible moment, fanning his tail as he does so, emitting a loud chirp noise.

Scientists, using high-speed video cameras established that the chirp is not emitted from his vocal organ, but from his tail. The outer tail feathers of the male Anna's hummingbird have a trailing van that vibrates as air passes through it at speed, creating the chirp sound, so the Picaflores are singing with their tails!

Peregrine falcons have been thought to be masters of the skies, but hummingbirds are the true owners of that title. Christopher Clark of the Museum of Vertebrate Zoology at Berkeley observed that at the point at which the hummingbirds pull up from their dive, their chirping tails fanned, and they experienced accelerations almost nine times greater than gravitational acceleration. These forces are the highest known for any vertebrate species undergoing a voluntary aerial maneuver with the exception of jet fighter pilots who have the benefit of special G-suits to counteract the considerable physiological effects of pulling 10Gs. Anna's hummingbirds have only a suit of feathers with a pleasing metallic raspberry-red gorget and head.

When courting, the male Anna's hummingbird ascends some 35 meters before diving over an interested female at equal to 385 body lengths per second, producing a high-pitched sound. In addition to acceleration, the speed, relative to body length is the highest known for any vertebrate, about twice the diving speed of peregrine falcons in pursuit of prey. At maximum descent speed, about 10 Gs of gravitational force occurs in the courting hummingbird during a dive. By comparison to humans, this is a G-force acceleration causing near loss of consciousness in fighter pilots during flight of fixed-wing aircraft in a high-speed banked turn.

The outer tail feathers of male Anna's and Allen's, calliope hummingbirds vibrate during courtship display dives and produce an audible chirp. Hummingbirds cannot make the courtship dive sound when missing their outer tail feathers. Picaflores can sing at the same frequency as the tail feather chirp, but it is not capable of the same volume. The chirp sound is caused by the aerodynamics of rapid air flow past tail feathers, causing them to flutter in a vibration which produces the high-pitched sound of a courtship dive.

Many other species of hummingbirds produce sounds with their wings or tails while flying, hovering or diving, including the wings of the calliope hummingbird, broad-tailed hummingbird, rufous hummingbird, Allen's hummingbird, and streamertail, as well as the tail of the Costa's hummingbird, the black-chinned hummingbird, and a number of related species. The harmonics of sounds during courtship dives vary across species of hummingbirds.

Male rufous and broad-tailed hummingbirds have a distinctive wing feature during normal flight that sounds like jingling or a buzzing shrill whistle that arises from air rushing through slots created by the tapered tips of the ninth and tenth primary wing feathers, creating a sound loud enough to be detected by female or competitive male hummingbirds and researchers up to 100 meters away.

The trill announces the sex and presence of a male bird and

provides audible aggressive defense of a feeding territory and an intrusion tactic. It also enhances communication of a threat and favors mate attraction and courtship.

Male Costa's hummingbirds twist their tail sideways to direct the sound of fluttering tail feathers toward a female when diving past her in courtship.

A SHARP DRESSED MAN

As an adjunct to courtship and territorial competition, many male hummingbirds have plumage with bright, varied coloration resulting both from pigmentation in the feathers and prismal cells within the top layers of feathers of the head, gorget, breast, back, and wings. When sunlight hits these cells it is split into wavelengths that reflect to the observer in varying degrees of intensity with the feather structure acting as a diffraction grating. Iridescent hummingbird colors result from a combination of refraction and pigmentation, since the diffraction structures themselves are made of melanin, a pigment, and may also be colored by carotenoid pigmentation with more subdued black, brown, or gray colors dependent on melanin.

By shifting position, feather regions of a muted-looking bird can instantly become fiery red or vivid green. In courtship displays males of the colorful Anna's hummingbird species orient their bodies and feathers toward the sun to enhance the display value of iridescent plumage toward a female of interest.

One study of Anna's hummingbirds found that dietary protein was an influential factor in feather color. Birds receiving more protein grew significantly more colorful crown feathers than those fed a low-protein diet, and birds on a high-protein diet grew yellower, higher hue green tail feathers than birds on a low-protein diet.

MASTERS OF THE AIR

Hummingbird flight has been studied intensively from an aerodynamic perspective using wind tunnels and high-speed video cameras. Two studies of rufous or Anna's hummingbirds in a wind tunnel used particle image velocimetry techniques to investigate the lift generated on the bird's up and downstrokes. The birds produced 75 percent of their weight support during the downstroke and 25 percent during the upstroke, with the wings making a "figure 8" motion.

A trail of wake vortices generated by a hummingbird's flight discovered after training a bird to fly through a cloud of helium-filled soap bubbles and recording airflows in the wake with stereo photography.

Many earlier studies assumed that lift was generated equally during the two phases of the wingbeat cycle, as is the case of insects of a similar size. This finding shows that hummingbird hovering is similar to, but distinct from that of hovering insects like the hawk moth. Further studies of hovering rufous hummingbirds showed that muscle strain in the pectoralis major, which is the principal downstroke muscle was the lowest recorded in a flying bird and the primary upstroke muscle is proportionately larger than in other bird species.

The giant hummingbird's wings beat as few as 12 beats per second and the wings of typical hummingbirds beat up to 80 times per second.

A slow-motion video has shown how hummingbirds deal with rain when they are flying. To remove the water from their heads, they shake their heads and bodies similar to a dog shaking, to shed water. Further, when raindrops collectively can weigh as much as 38 percent of the bird's body weight, hummingbirds shift their bodies and tails horizontally, beat their wings faster, and reduce their wings' angle of motion when flying.

THE EVOLUTION OF THE FAMILY TREE

A map of the hummingbird family tree reconstructed from 284 of the world's 338 known species shows rapid diversification from 22 million years ago. Hummingbirds fall into nine main subdivisions, the topazes, hermits, mangoes, brilliants, coquettes, patagona, mountain gems, bees, and emeralds, defining their relationship to nectar-bearing flowering plants and the birds' continued spread into new geographic areas.

While hummingbirds depend on flower nectar to fuel their high metabolisms and hovering flight, changes in flower and beak shape stimulated the formation of new species of hummingbirds and plants. Due to this exceptional evolutionary pattern as many as 140 hummingbird species can coexist in a specific region like in the Andes.

The hummingbird family tree shows ancestral hummingbirds splitting from insectivorous swifts and treeswifts about 42 million years ago, probably in Eurasia. One key evolutionary factor appears to be an altered taste receptor that enabled them to seek nectar. By 22 million years ago the ancestral species of current hummingbirds became established in South America, where environmental conditions stimulated further diversification.

The Andes appear to be a particularly rich environment for hummingbird evolution because diversification occurred simultaneously with mountain uplift over the past 10 million years. Hummingbirds remain in dynamic diversification inhabiting ecological regions across South America, North America, and the Caribbean, indicating an enlarging evolutionary expansion. Within the same geographic region, hummingbird evolution from ancestors co-evolved with nectar-bearing plant evolution, affecting mechanisms of pollination.

Hummingbirds exhibit sexual size dimorphism according to Rensch's rule in which males are smaller than females in small species, and males are larger than females in large-bodied species. The extent of this sexual size difference varies among evolutionary subdivisions of hummingbirds. While the Mellisugini subdivision exhibits a large size dimorphism with females being larger than males, the Lophomithini category displays little size dimorphism, so males and females are similar in size. Sexual dimorphisms in beak size and shape are also present between male and female hummingbirds. In many subdivisions females have longer, more curved beaks favored for accessing nectar from tall flowers. In the case of males and females of the same size, females tend to have larger beaks.

Sexual size and beak differences likely evolved due to constraints imposed by courtship because mating displays of male hummingbirds requires complex aerial maneuvers. Males tend to be smaller than females, allowing conservation of energy to forage competitively and participate more frequently in courtship, so sexual selection favors smaller male hummingbirds.

Female hummingbirds tend to be larger, requiring more energy, with longer beaks that allow for more effective reach into crevices of tall flowers for nectar, so females are better at foraging, acquiring flower nectar, and supporting the energy demands of their larger body size. Directional selection favors larger hummingbirds in terms of acquiring food.

Another evolutionary cause of this sexual beak dimorphism is the selective forces from competition for nectar between the sexes of each species that drive the sexual dimorphism. Depending on which sex holds territory in the species, it is advantageous for the other sex to have a longer beak to be able to feed on a wide variety of flowers, decreasing competition.

In species of hummingbirds where males have longer beaks, males do not hold a specific territory and have male courtship displays that define their mating system. In species where males have shorter beaks than females, males defend their resources, so females have a longer beak to feed from a broader range of flowers.

Hummingbirds are specialized nectarivores and are tied to the bird pollinated flowers they feed on. Some species, especially those with unusual beak shapes like the sword-billed hummingbird and the sicklebills co-evolved with a small number of flower species, but even in the most specialized hummingbird-plant relationships the number of food plant lineages of individual hummingbird species increases with time. The bee hummingbird which is the world's smallest bird evolved to dwarfism most likely because it had to compete with long-billed hummingbirds having an advantage for nectar foraging from specialized flowers, leading the bee hummingbird to compete more successfully for flower foraging against insects.

Many plants pollinated by hummingbirds produce flowers in shades of red, orange, and bright pink, though the birds will take nectar from flowers of other colors. Hummingbirds can see wavelengths into the near-ultraviolet, but hummingbird-pollinated flowers do not reflect these wavelengths like insect-pollinated flowers do. This narrow color spectrum may render hummingbird-pollinated flowers inconspicuous to most insects, reducing nectar robbing. Hummingbird-pollinated flowers also produce weak nectar, while insect-pollinated flowers typically produce more concentrated nectars dominated by fructose and glucose.

Mating season is competitive for hummingbirds. After a little bobbing and weaving, territorial males use their needle-like beaks like tiny knives and stab each other in the throat. They also use their face-knives for hunting. Nectar is their favorite food source, but not the only one. Picaflores also eat small flying insects. When they approach a bug, they stretch their beaks wide until the beak reaches its maximum stretch, then it snaps shut like a hair clip, trapping the insect inside.

In traditional taxonomy hummingbirds are placed in the order Apodiformes which also contains the swifts, but some taxonomists have separated them into their own order, the Trochiliformes. Hummingbirds' wing bones are hollow and fragile, making fossilization difficult so their evolutionary history is poorly documented. Though scientists theorize that hummingbirds originated in South America where species diversity is greatest, their ancestors may have lived in parts of Europe to southern Russia.

Between 325 and 340 species of hummingbirds are documented, divided into two subfamilies, the hermits, consisting of 34 species in six genera, and the typical hummingbird subfamily Trochilinae. Recent analyses suggest that this division is inaccurate, and that there are nine major subfamilies of hummingbirds: the topazes and jacobins, the hermits, the mangoes, the coquettes, the brilliants, the giant hummingbird, the mountaingems, the bees, and the emeralds. The topazes and jacobins combined have the oldest split with the rest of the hummingbirds. The hummingbird family has the second-greatest number of species of any bird family after the tyrant flycatchers.

Dr. Gerald Mayr of the Senckenberg Museum in Frankfurt am Main identified two 30-million-year-old hummingbird fossils of a primitive hummingbird species named *Eurotrochilus inexpectatus*, which means unexpected European hummingbird, had been sitting in a museum drawer. They had been unearthed in a clay pit at Wiesloch–Frauenweiler, south of Heidelberg, Germany, and because it was assumed that hummingbirds never occurred outside the Americas, these were not recognized as hummingbirds until Mayr took a closer look at them.

Fossils of birds not clearly assignable to either hummingbirds or a related extinct family, the Jungornithidae, have been found at the Messel pit and in the Caucasus, dating from 40–35 million years ago indicating that the split between these two lineages occurred at that date. The locations where these early fossils were found both had climates similar to those found in the northern Caribbean and southern China during that time. The biggest remaining mystery at present is what happened to hummingbirds in the roughly 25 million years between the primitive *Eurotrochilus* and the modern fossils. The astounding morphological adaptations, the decrease in size, and the dispersal to the Americas and extinction in Eurasia all occurred during this period. DNA-DNA hybridization results suggest that the main radiation of South American hummingbirds took place at least partly in the Miocene, some 12 to 13 million years ago during the uplifting of the northern Andes. In 2013, a 50 million year old bird fossil discovered in Wyoming was determined to be a predecessor to both hummingbirds and swifts before the groups diverged.

FUEL EFFICIENCY

No other bird on Earth can fly like a hummingbird. They can fly forward or backward, hover, and even fly upside-down, and they do all of this so fast we can't even see the beating of their wings between 70 and 200 times per second. This power, precision, and agility allows them to reach speeds of up to 30 miles per hour while flying and 60 miles per hour while diving, but like marathon runners, hummingbirds have super-fast metabolisms and need to eat about every 10 minutes. It's believed that they eat two to three times their own body weight in bugs and nectar every day, the human equivalent of an entire refrigerator full of food.

With the exception of insects, hummingbirds in flight have the highest metabolism of all animals which is necessary to support the rapid beating of their wings during hovering and fast forward flight. Their heart rate can reach as high as 1,260 beats per minute with a breathing rate of 250 breaths per minute, even at rest. During flight, oxygen consumption per gram of muscle tissue in a hummingbird is about 10 times higher than that measured in top performing human athletes.

Hummingbirds are rare among vertebrates in their ability to rapidly use ingested sugars to fuel energetically expensive hovering flight, powering up to 100 percent of their metabolic needs with the sugars they drink while human athletes max out at around 30 percent. Hummingbirds can use newly-ingested sugars to fuel hovering flight within 30–45 minutes of consumption. The data indicates that hummingbirds can oxidize sugar in flight muscles at rates high enough to satisfy their extreme metabolic demands to support their high metabolic rate for hovering, foraging at altitude, and migrating.

By relying on newly-ingested sugars to fuel flight, hummingbirds can reserve their limited fat stores to sustain overnight fasting or power migratory flights. Studies of hummingbird metabolism address how a migrating ruby-throated hummingbird can cross 500 miles of the Gulf of Mexico on a nonstop flight. This hummingbird, like other long-distance migrating birds, stores fat as a fuel reserve augmenting its weight by as much as 100 percent, enabling metabolic fuel for flying over open water.

The high metabolic rate of hummingbirds, especially during rapid forward flight and hovering produces increased body heat that requires specialized mechanisms for heat dissipation, which becomes a greater challenge in hot, humid climates. Hummingbirds dissipate heat partially by evaporation through exhaled air, and from body structures with thin or no feather covering, such as around the eyes, shoulders, under the wings, and feet.

While hovering, hummingbirds do not benefit from heat loss by air convection during forward flight, except for air movement generated by their rapid wing-beat, possibly aiding convective heat loss from their extended feet. Smaller hummingbird species, such as the calliope, appear to adapt their relatively higher surface-to-volume ratio to improve convective cooling from air movement by their wings. When air temperatures rise above 97 degrees Fahrenheit, thermal gradients driving heat passively by convective dissipation from around the eyes, shoulders, and feet are reduced or eliminated, requiring heat dissipation mainly by evaporation and exhalation. In cold climates, hummingbirds retract their feet into breast feathers to eliminate skin exposure and minimize heat dissipation.

The dynamic range of metabolic rates in hummingbirds requires a parallel dynamic range in kidney function. During a day of nectar consumption with a corresponding high water intake that may total five times the body weight per day, hummingbird kidneys process water in amounts proportional to water consumption, avoiding overhydration. During brief periods of water deprivation such as what occurs in nighttime torpor, filtration drops to zero, preserving body water.

Hummingbird kidneys also have a unique ability to control the levels of electrolytes after consuming nectars with high amounts of sodium and chloride or none, indicating that kidney structures have evolved to be highly specialized to adjust for variations in nectar and mineral quality. Morphological studies which have been done on Anna's hummingbird kidneys showed adaptations of high capillary density, allowing precise regulation of water and electrolytes.

During turbulent airflow conditions created in a wind tunnel, hummingbirds exhibit stable head positions and orientation when they hover at a feeder. When wind gusts from the side, hummingbirds compensate by increasing wing-stroke amplitude and stroke plane angle by varying these parameters asymmetrically between the wings from one stroke to the next. They also vary the orientation and enlarge the collective surface area of their tail feathers into the shape of a fan. While hovering, the visual system of a hummingbird is able to separate apparent motion caused by the movement of the hummingbird itself from motions caused by external sources like an approaching predator. In natural settings full of highly complex background motion, hummingbirds can hover in place by precise rapid coordination of vision with body position.

The metabolism of hummingbirds slows at night or when food is not available. Picaflores enter a hibernatory, deep-sleep state known as torpor to prevent energy reserves from falling to a critical level. During nighttime torpor, body temperature falls from 104 to 64 degrees Fahrenheit, with heart and breathing rates slowed dramatically to roughly 50 to 180 beats per minute from its daytime rate of higher than 1000.

To prevent dehydration during torpor kidney filtration ceases, preserving needed compounds like glucose, water, and nutrients. Body mass declines throughout nocturnal torpor at a rate of 0.04 grams per hour, amounting to about 10 percent of weight loss each night. The circulating hormone, corticosterone, is one of the signals that arouses a hummingbird from torpor.

Use and duration of torpor varies among hummingbird species and are affected by whether a dominant bird defends territory with nonterritorial subordinate birds having longer periods of torpor.

Picaflores have long lifespans for organisms with such rapid metabolisms, but many die during their first year of life, especially in the vulnerable period between hatching and fledging. Those that survive can live a decade or more. Among the better known North American species, the average lifespan is about 3 to 5 years. For comparison, the smaller shrews, among the smallest of all mammals, seldom live longer than 2 years. The longest recorded lifespan in the wild belongs to a female broad-tailed hummingbird that was banded as an adult at least one year old and recaptured 11 years later, making her at least 12 years old. Other longevity records for banded hummingbirds include an estimated minimum age of 10 years 1 month for a female black-chinned hummingbird similar in size to the broad-tailed hummingbird, and at least 11 years 2 months for a much larger buff-bellied hummingbird.

HOME IS WHERE THE HEART IS

As far as is known male hummingbirds do not take part in nesting. Most species build a cup-shaped nest on the branch of a tree or shrub, although a few tropical species attach their nests to leaves. Nests vary in size relative to the particular species – from smaller than half a walnut shell to several centimeters in diameter.

Many Picaflores use spider silk and lichen to bind the nest material together and secure the structure. The unique properties of the silk allow the nest to expand as the young hummingbirds grow. Two white eggs are laid, which despite being the smallest of all bird eggs are large relative to the adult hummingbird's size. Incubation lasts 14 to 23 days, depending on the species, ambient temperature, and female attentiveness to the nest. The mother feeds her nestlings on small arthropods and nectar by inserting her beak into the open mouth of a nestling and regurgitating the food into its crop. Newborn hummingbirds stay in the nest for 18 to 22 days, then leave it to forage on their own, but their mother may continue feeding them for another 25 days.

In present times Picaflores only live within the Americas from south central Alaska to Tierra del Fuego, including the Caribbean. The majority of species live in tropical and subtropical Central and South America, but several species also breed in temperate climates and some hillstars live in alpine Andean highlands at altitudes up to 17,100 feet.

The greatest species richness is in humid tropical and subtropical forests of the northern Andes and adjacent foothills, but the number of species found in the Atlantic Forest, Central America, or southern Mexico exceeds the numbers found in southern South America, the Caribbean islands, the United States, and Canada. While fewer than 25 different species of hummingbirds have been recorded from the United States and fewer than 10 from Canada and Chile. Colombia has more than 160 and the comparably small Ecuador has about 130.

The migratory ruby-throated hummingbird breeds in a range from the southeastern United States to Ontario, while the black-chinned hummingbird, its close relative and another migrant, is the most widespread and common species in the southwestern United States. The rufous hummingbird is the most widespread species in western North America and the only hummingbird documented outside of the Americas in the Chukchi Peninsula of Russia.

Most North American hummingbirds migrate south in the fall to spend winter in Mexico, the Caribbean Islands, or Central America. Some southern South American species also move north to the tropics during the southern winter. A few species are year-round residents of Florida, California, and the far southwestern desert regions of the United States. Among these are Anna's hummingbird, a common resident from southern Arizona and inland California, and the buff-bellied hummingbird, a winter resident from Florida across the Gulf Coast to southern Texas. Ruby-throated hummingbirds are common along the Atlantic flyway and migrate in summer from as far north as Atlantic Canada, returning to Mexico, South America, southern Texas, and Florida to winter. In southern Louisiana, black-chinned, buff-bellied, calliope, Allen's, Anna's, ruby-throated, rufous, broad-tailed, and broad-billed hummingbirds are present during winter months.

The rufous hummingbird breeds farther north than any other species, breeding in large numbers in temperate North America and wintering in increasing numbers along the coasts of the subtropical Gulf of Mexico and Florida instead of western or central Mexico. By migrating in spring as far north as the Yukon or southern Alaska, the rufous hummingbird migrates more extensively and nests farther north than any other hummingbird species, and has to tolerate occasional temperatures below freezing in its breeding territory. This cold hardiness enables it to survive temperatures below freezing, as long as adequate shelter and food are available.

As calculated by displacement of body size, the rufous hummingbird makes what is most likely the longest migratory journey of any bird in the world. At just over 3 inches long, they travel 3,900 miles one-way from Alaska to Mexico in late summer, a distance equal to 78,470,000 body lengths. By comparison, the 13-inch-long Arctic tern makes a one-way flight of about 11,185 miles, or 51,430,000 body lengths, just 65 percent of the body displacement during migration by rufous hummingbirds. The northward migration of rufous hummingbirds occurs along the Pacific flyway and may be time-coordinated with flower and tree leaf emergence in spring in early March, and the availability of insects as food. Arrival at breeding grounds before nectar availability from mature flowers can jeopardize breeding opportunities.

A BEAK EXPERIENCE

For nutrition, hummingbirds eat a wide variety of insects, including mosquitoes, fruit flies, gnats in flight or aphids on leaves and spiders in their webs. Their lower beak is flexible and can bend as much as 25 degrees when it widens at the base, making a larger surface for catching insects. Hummingbirds hover within insect swarms in a method called "hover-hawking", a feeding strategy involving catching flying insects in the air. The term usually refers to a technique of sallying out from a perch to snatch an insect and then returning to the same or a different perch, though it also applies to birds that spend most of their lives on the wing. The term "hawking" comes from the similarity of this behavior to the way hawks take prey in flight, although raptors catch prey with their feet, hawking is the behavior of catching insects in the beak.

To supply energy needs, hummingbirds drink nectar from inside flowers. Like bees, they are able to assess the amount of sugar in the nectar they drink and reject flower types that produce nectar that is less than 10 percent sugar, preferring those with higher sugar content. Nectar is a mixture of glucose, fructose, and sucrose, and is a poor source of nutrients, requiring hummingbirds to meet their nutritional needs by consuming insects.

Picaflores do not spend all day flying, as the energy cost is prohibitive and the majority of their activity consists of sitting or perching. They eat many small meals and consume around half their weight in nectar or twice their weight in nectar if it is 25 percent sugar each day. Hummingbirds digest their food rapidly due to their small size and high metabolism with a mean retention time of less than an hour. Hummingbirds spend an average of 10–15 percent of their time feeding and 75–80 percent sitting and digesting.

Because their high metabolism makes them vulnerable to starvation, hummingbirds are highly attuned to food sources. Some species, including many in North America are territorial and will guard food sources like feeders against other hummingbirds, attempting to ensure a future food supply for itself. Hummingbirds also have an enlarged hippocampus, the brain region that facilitates spatial memory used to map flowers previously visited during nectar foraging.

Hummingbird beaks are flexible and their shapes vary dramatically as an adaptation for specialized feeding. Some species like hermits have long beaks that allow them to probe deep into flowers with long corollae. Thornbills have short, sharp beaks adapted for feeding from flowers with short corollae and piercing the bases of longer ones. The sicklebills' extremely decurved beaks are adapted to extracting nectar from the curved corollae of flowers in the family Gesneriaceae and the beak of the fiery-tailed Awlbill has an upturned tip like the avocets. The male tooth-billed hummingbird has barracudalike spikes at the tip of its long, straight beak.

The two halves of a hummingbird's beak have a pronounced overlap, with the lower mandible fitting tightly inside the upper maxilla. When a hummingbird feeds on nectar, the beak is usually opened only slightly, allowing the tongue to dart out and into the interior of flowers. Hummingbird beak sizes range from about 5 millimeters to as long as 100 millimeters which is about 4 inches. When catching insects in flight, a hummingbird's jaw flexes downward to widen their beak for successful capture.

Perception of sweetness in nectar evolved in hummingbirds during their genetic divergence from insectivorous swifts, their closest bird relatives. Although the only known sweet sensory receptor, called T1R2, is absent in birds, receptor expression studies showed that hummingbirds adapted a carbohydrate receptor from the T1R1-

T1R3 receptor, identical to the one perceived as umami in humans, repurposing it to function as a nectar sweetness receptor. This adaptation for taste enabled hummingbirds to detect and exploit sweet nectar as an energy source, facilitating their distribution across geographical regions where nectar-bearing flowers are available.

For years scientists believed that Picaflores used their tongues like tiny straws to suck up nectar until researchers discovered that they have forked tongues lined with fine hair-like extensions called lamellae. When the hummingbird starts drinking, the tongue's forks open, the lamellae unroll and curl around a drop of nectar, then as the tongue is brought back into the mouth, the forks close and the lamellae trap the nectar.

A hummingbird's tongue is translucent and thin as embroidery thread. At a feeder, a hummingbird extrudes and withdraws the tongue 13 times a second, but they do not sip nectar, they lap it. The tongue is forked, like a snake's, with absorbent fringes along the edge of each fork, and grooved down the center to withdraw extra nectar through capillary action. The tongue is so long that when retracted, it extends back to the rear of the skull and curls around on top of the skull. With this extraordinary appendage, a hummingbird can drink its own weight in nectar in a single visit to a feeder.

In the wild, hummingbirds visit flowers for food, extracting nectar, which is 55 percent sucrose, 24 percent glucose and 21 percent fructose on a dry-matter basis and they take sugar-water from bird feeders. A drawback of artificial feeders is that the birds seek less flower nectar for food, which reduces the amount of pollination their feeding naturally provides.

White granulated sugar is the best sweetener to use in hummingbird feeders, using 1 part sugar to 4 parts water, although hummingbirds will defend feeders more aggressively when sugar content is at 35 percent, indicating preference for nectar with a higher sugar content. Organic and "raw" sugars contain iron, which can be harmful, and brown sugar agave syrup, molasses and artificial sweeteners also should not be used. Honey is made by bees from the nectar of flowers, but it is not good to use in feeders because when diluted with water, microorganisms easily grow in it, causing it to spoil rapidly.

Red food dye was once thought to be a favorable ingredient for feeders, but there is no point in adding it to the nectar. Some people speculate red dye could be bad for the birds, although this claim has not received scientific attention. Commercial products sold as "instant nectar" or "hummingbird food" can contain preservatives and artificial flavors as well as dyes. Although some commercial products contain small amounts of nutritional additives, hummingbirds obtain all necessary nutrients from the insects they eat, rendering added nutrients practically useless.

Bees, wasps, ants, and other animals attracted to the sugar-water also visit hummingbird feeders and can crawl into them and become trapped and drowned. Orioles, woodpeckers, bananaquits, raccoons and other larger animals are also known to drink from hummingbird feeders, sometimes tipping them and draining the liquid. In the southwestern United States, two species of nectar-drinking bats visit hummingbird feeders to supplement their natural diet of nectar and pollen from saguaro cacti and agaves.

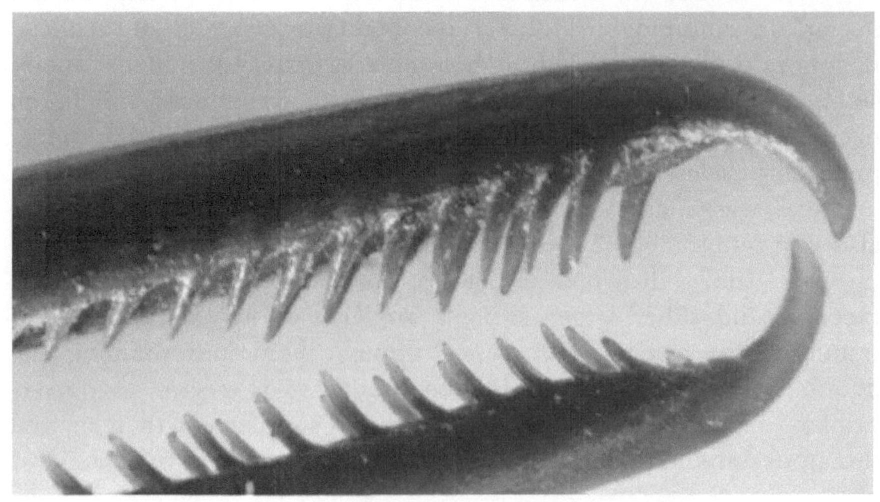

A hummingbird beak tip close up

An extended hummingbird tongue

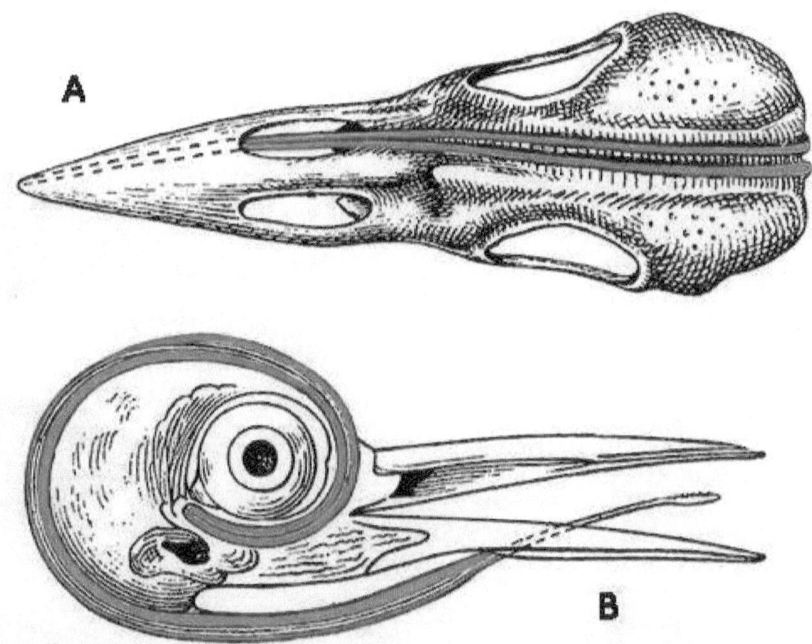

A hummingbird's tongue wraps around its eyes inside its head and has a forked tip that grabs nectar, rather than sucking it up through capillary muscular action.

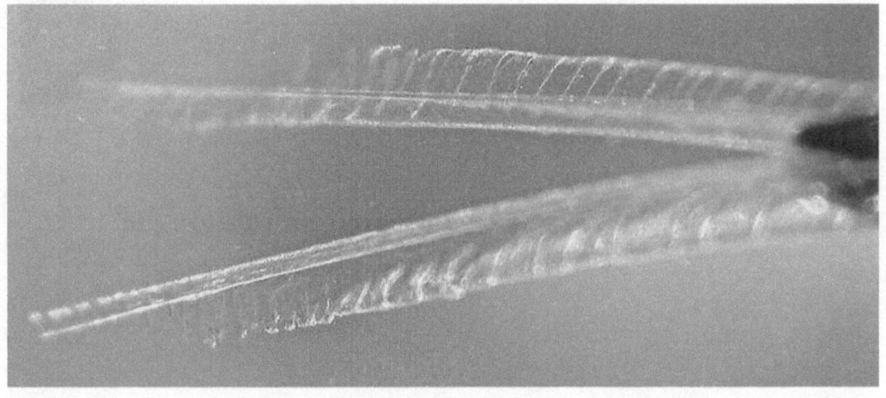

A Hummingbird tongue close up

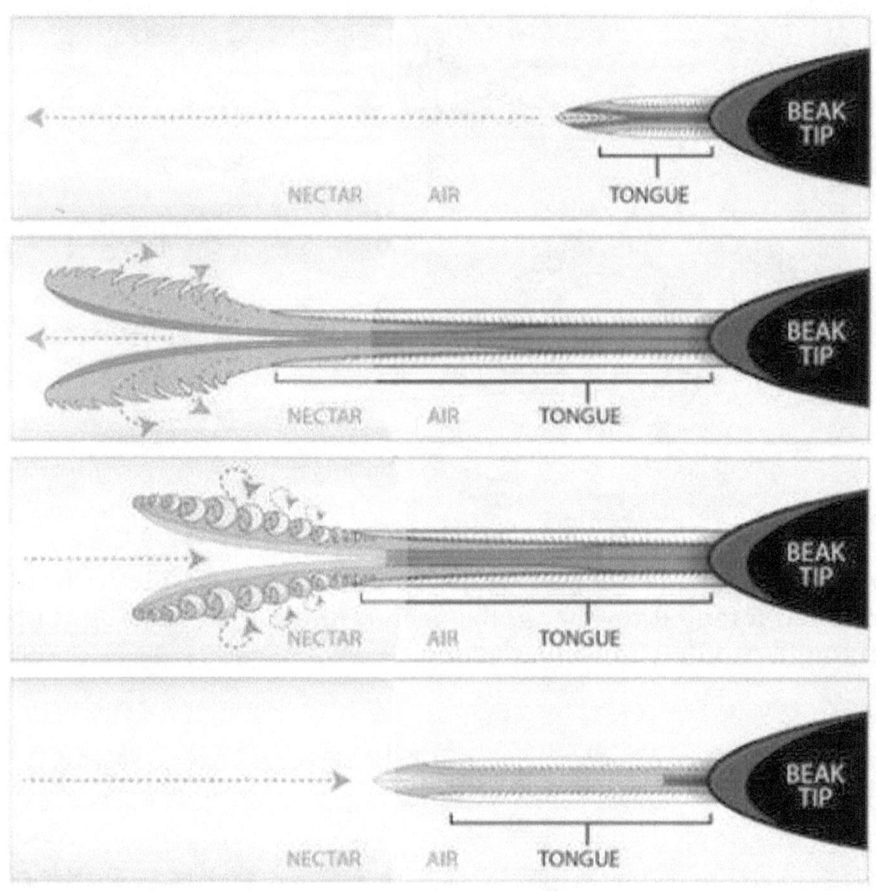

The lapping action of a hummingbird's tongue

THE LIGHT THAT HOVERS
BETWEEN SPIRIT AND MATTER

The high frequency neon pastel visions I have experienced communing with Picaflor are beyond description and unlike *anything* I have ever seen in the natural world. Humans are virtually color blind compared to birds and many other animals. Picaflores far outperform humans in distinguishing hues in combinations of ultraviolet and visible light which is critically important to keeping the ecosystem in balance in a world loaded with beautiful flowers and fruit. A hummingbird's vision aids their ability to identity certain flowers and ramp up pollination.

Picaflores have an advantage with their vision because they possess a fourth color cone that can detect ultraviolet light. Humans have just three color cones in their retina that are sensitive to red, green, and blue light. This ability helps hummingbirds and many of their feathered friends deliver on the prime ecological functions of pollination by sight helping Picaflores locate nectar-bearing flowers covered in patterns that are imperceptible to humans.

On top of that, hummingbirds are among the many animals

gifted with a third set of eyelids. These translucent flaps of skin known as nictitating membranes act like natural flight goggles, protecting the hummingbird's eyes as they zoom through the air.

Hummingbirds have exceptional visual acuity that provides them with discrimination of food sources while foraging. It is commonly believed that hummingbirds are attracted to color while seeking food like red flowers or feeders. Research has shown that location and flower nectar quality are the most important "beacons" for foraging. Hummingbirds didn't depend much on visual cues of flower color to locate nectar rich locations, they used surrounding landmarks to find the nectar.

In at least one hummingbird species, the green-backed firecrown, preferred flower colors are in the red-green wavelength for the bird's visual system, providing a higher contrast than other flower colors. Additionally, the crown plumage of firecrown males is highly iridescent in the red wavelength range, possibly providing a competitive advantage of dominance when foraging among other hummingbird species with less colorful plumage. The ability to discriminate colors of flowers and plumage is enabled by a visual system that has four single cone cells and a double cone screened by photoreceptor oil droplets that enhance color discrimination.

During evolution hummingbirds have adapted to the navigational needs of visual processing while in rapid flight or hovering by development of an exceptionally dense array of retinal neurons that allow increased spatial resolution in the lateral and frontal visual fields. Studies have shown that neuronal hypertrophy, relatively the largest in any bird, exists in a brain region called the pretectal nucleus lentiformis mesencephali, which is the nucleus of the optic tract in mammals responsible for refining dynamic visual processing while hovering and in rapid flight. The enlargement of this brain region responsible for visual processing indicates an enhanced ability for perception and processing of fast-moving visual stimuli which hummingbirds encounter during rapid forward flight, insect foraging, competitive interactions, and high-speed courtship.

Hummingbirds are highly sensitive to stimuli in their visual fields and respond to minimal motion in any direction by reorienting themselves in midflight. Their visual sensitivity allows them to precisely hover in place while in complex and dynamic natural environments, functions enabled by the lentiform nucleus which is tuned to fast-pattern velocities that enable highly tuned control and collision avoidance during forward flight.

Hummingbirds love the color red because their eyes are tuned to the rosy hue with retinas that have a denser concentration of cones that mute color shades like blue and heighten warmer shades like red and yellow.

Picaflores' large eyes pick up even more colors than human eyes with retinas that possess four types of cones. Humans only have three, which detect blue, green, and red light. This fourth type of cone is sensitive to ultraviolet or UV light, which human eyes cannot see, and though an admittedly subjective experience, this describes the brilliant luminescent colors I have seen in my visionary experiences when I have communed with Picaflor that I have never seen in everyday life.

Hummingbirds see the world in a range of brilliant colors that we can only dream of or see in visions. When the UV hues blend with the ones that humans can perceive, new colors appear. This visual acuity also appears in other birds, reptiles, and fish, and likely helps them locate food, choose a mate, and elude predators.

We see red combined with blue as purple, but birds can see purple and ultraviolet. In normal human perception, the phrase "every color of the rainbow" isn't as all-encompassing as it sounds because the color purple is missing from rainbows.

The "V" in "R O Y G B I V" stands for violet, but violet is not purple. There is no purple wavelength of light. It requires a mixture of both red and blue wavelengths making it a non-spectral color, and is in fact the only non-spectral color humans see. It requires our brains to interpret signals from both red-sensitive and blue-sensitive cones in our eyes in order to see purple as a separate color.

While humans have three types of cones that make us trichromatic, many creatures have four, expanding their visible spectrum into ultraviolet wavelengths that mix with red, yellow, green, and purple, indicating that they can see additional non-spectral colors that humans struggle to imagine in normal modes of perception outside of visionary experiences.

Research on bees has demonstrated that they see UV and green as its own color, referred to as bee-purple, but there isn't a lot of experimental evidence beyond that. A team led by Princeton's Mary Stoddard decided to test the idea by taking advantage of hummingbirds' love of sugar-water feeders.

Researchers set up a pair of feeders for their experiments, one containing sugar water and one containing plain water. On top of each they attached an LED light containing UV, blue, red, and green LEDs behind a diffuser, allowing the researchers to light up the feeder in a variety of non-spectral colors.

They watched wild broad-tailed hummingbirds come to visit, recording which feeder they flew up to first. After a number of visits, the feeder positions were switched so the birds couldn't return to the same spot once they found the sugar water. The idea was that they would use the color of the light to identify the feeder on return visits. The researchers couldn't track individual birds separately, but based on some banding, they estimated the local population at 200 to 300 and recorded over 6,000 hummingbird visits.

The experiments pitted different pairs of colors together coupled with a few control runs where both lights displayed the exact same color and some experiments testing red vs. green. From there, the differences became more subtle and depended on differentiating non-spectral colors. Most involved different mixtures of UV and another color in the same way that we differentiate between reddish-purple and a bluish-purple.

The tests showed that the birds could see every non-spectral color that the researchers threw at them. Color pairs closer together in hue resulted in more mistaken visits but still beat the odds of the control experiments.

As an additional plausibility check, researchers scanned databases of precisely measured colors that appear in plants and birds. These non-spectral colors are common in nature, accounting for 30 percent of bird plumage colors and 35 percent of plant colors in the databases. It made sense that hummingbirds and other birds can see those colors in their environment.

The researchers believe that this study is generalizable beyond the broad-tailed hummingbirds that volunteered for it. Many things are poorly understood about the physiology of eyesight across bird species, much less the neural processing of signals from the color cones in the eye.

"Although these experiments were performed with hummingbirds," the team writes, "our findings are likely relevant to all diurnal, tetrachromatic birds and probably to many fish, reptiles, and invertebrates."

"Even if the neural mechanisms for color vision were clear, and even if color-mixing experiments attest to avian tetrachromacy," they write, "we still could not answer the more philosophical question of what non-spectral colors really look like to birds. Does UV + green appear to birds as a mix of those colors (analogous to a double-stop chord played by a violinist) or as a sublime new color (analogous to a completely new tone unlike its components)? We cannot say."

Whether real or imagined, I personally have been blessed to experience what this is like in my visions with Picaflor.

A PATH WITH HEART

Our head is the center of perception for four of our senses. Our fifth sense is centered in the body, our primary organ of feeling, and at the center of that is our heart which is the center of feeling and emotion. Ancient shamanic cultures saw the heart as the true center of their personal universe, functioning as the portal connecting them through the sun of our local solar system through multiple permutations all the way back to the infinite source of the great mystery.

Our arrival at our center brings us to our tactile senses that literally cover our whole body with greater concentrations of nerve endings located at critical points like our fingertips and face. This multifaceted and pervasive network is a complex extended system of sensory neurons and pathways known as our somatosensory system that responds to changes at the surface or inside our bodies. The axon nerve fibers of sensory neurons connect and respond to various sensory receptor cells that are activated by stimuli like heat and pain, giving a functional name to the responding sensory neurons like thermoreceptors that carry information about temperature changes. These nerve endings are vibratory sensors of heat, a form of energy produced when molecules increase their rate of oscillation which is a raising of frequency.

In many ways we experience our psyches as outside, but using our dreams as an example, if they were strictly internal, they would take us wandering around inside the organs and tissues of our own bodies, but they do not. They take us to tropical islands, strange nether worlds, and the sky as well as many other places.

From our subjective experience our dreams are "out there" in the cosmos of dreaming. If we remain faithful to our experience the way indigenous societies have, we have to say that the unconscious of our dreaming sojourns are not invisible realms inside our heads, hearts, or stomachs. The domains that we travel through in our dreams, shamanic journeys, and visions are experienced as an alternate cosmos with a different set of rules, invisible to our ordinary awareness of space and time while suffusing and extending it far beyond in infinite directions.

If we try to give an account of a nonrational experience that attempts to describe things the way we experienced them, for example shamanic journeys that take place in the unconscious, what we mean is that we travel through a realm that is unknown to ordinary awareness, one that does not take place in the space and time of our everyday experience of consensual reality. It takes place in an *imaginal cosmos* no less real than this one, but radically different.

Furthermore, like the hypothesis of the unconscious, the shamanic realm of imaginal sojourns is in a sense *more real* than that of everyday awareness, making psychology's definition of the unconscious the greater reality within which our ordinary awareness is too fragmentary and narrow to adequately understand itself. Here, the shamanic point of view is in agreement. Shamanism finds that everyday events have a larger meaning that can only be appreciated when we journey out of the everyday into the greater cosmos that encompasses this little one, so the forces of the shamanic cosmos shape and determine what happens to us in our everyday lives.

All of this interconnectedness and interaction reinforces the notion that we are all one, connected in our hearts, minds, and spirits. Shamans believe that everything is interconnected, including man, plants, animals, trees, and for that matter every part of creation. American Indians often refer to these other beings as brothers.

On an esoteric level, everything is connected within everything else in what modern physicists call a holographic manner. To record a hologram, a coherent light beam passes through a beam splitter. Some of the light scattered from an object or a set of objects falls on the recording medium. The other part of the split beam reflects off of a mirror. This second light beam is known as the reference beam, which also illuminates the recording medium, so that interference occurs between the two beams. The resulting light field generates a seemingly random pattern of varying intensity, which is recorded in the hologram.

An interesting property of holograms is that if you cut one up into smaller pieces, each portion contains all of the information about the whole object.

In shamanism the heart in man is the sun of his personal cosmos which is connected to the sun, which is the giver of life in our solar system, especially on planet earth. This sun in turn is connected to a bigger sun, which is connected to a bigger sun, infinitely, all the way back to Source.

Hafez, a 14th-century poet, best known for expressing the ecstasy of divine inspiration in the mystical form of love poems penned a wonderful verse that captures the essence of this belief.

Even After All this time The Sun never says to the Earth, "You owe me." Look What happens With a love like that, It lights the whole sky.

An old adage in shamanism made popular by Carlos Castaneda attributed to Don Juan, the Yaqui shaman, is that, "A warrior must follow a path with heart." In other words, follow your heart back to Source and you will have found your way home.

Manifesting as air wrapped in light, every aspect of the physical and essence energetic of Picaflor is high frequency, not only in their external displays of high frequency color, song, high speed wing hum, and feather trills from their tails, but internally in how they see, hear, metabolize, and burn energy at high frequencies. Other than their bigger hearts, huge lungs, and brains, they are mostly made from air, the elemental defined as and representative of spirit, and they are masters of air. If they existed and manifested at frequencies that are higher than they already are, they would disappear into invisibility beyond our perception, in essence becoming pure spirit.

Is it any wonder that Picaflores are considered to be the nerve endings of God?

Crop circle on Milk Hill, Nr Stanton St Bernard, Wiltshire in England that appeared on July 2, 2009

BIBLIOGRAPHY

Dunn, John, *The Glitter in the Green: In Search of Hummingbirds*, (*New York: Basic Books, 2021*)

Montgomery, Sy, *The Hummingbird's Gift*, (New York: Atria Books, 2021)

Pallamary, Matthew J., *Picaflor,* (San Diego: Mystic Ink Publishing, 2021)

Pallamary, Matthew J., *Spirit Matters,* (San Diego: Mystic Ink Publishing, 2007)

Pallamary, Matthew J., *The Center Of The Universe Is Right Between Your Eyes But Home Is Where the Heart Is,* (San Diego: Mystic Ink Publishing, 2017)

Prechtel, Martin, *The Disobedience of the Daughter of the Sun*, (Berkeley, 2005)

http://advances.sciencemag.org/cgi/content/full/6/29/eabb9393/DC1

https://advances.sciencemag.org/content/6/29/eabb9393?fbclid=IwAR1eL_3Coi7DehgPFz-i5sPYwFjHxH7h7c6sax3Iu-e-GbZRGkQZUIvyciE

https://aneace.com/blog/the-eagle-and-the-hummingbirds

https://www.animalwised.com/the-mayan-hummingbird-legend-gift-to-the-hummingbird-2012.html

https://apnews.com/2c9fa3b879038da9d2c04e3b650630f3

https://www.audubon.org/news/the-origins-hummingbirds-are-still-major-mystery

https://bgr.com/2021/03/17/why-hummingbirds-hum-study/?bgr-partner=flipboard

https://www.birdsandblooms.com/birding/attracting-hummingbirds/13-questions-hummingbird-feeders-answered-pros/

https://birdwatchingbuzz.com/hummingbird-symbolism/

https://cedarhilllonghouse.ca/the-hummingbird-symbol-native-art-symbols-and-meanings/

https://www.crystalinks.com/hummingbird.html

https://fb.watch/4Eg-7XVaxG/

https://www.firstpeople.us/FP-Html-Legends/Gift_To_The_Hummingbird-Mayan.html

https://www.google.com/search?q=hummingbird+brain&rlz=1C1WNCD_enUS895US895&oq=hummingbird+bra&aqs=chrome.0.0l3j69i57j0.10267j0j7&sourceid=chrome&ie=UTF-8

https://www.google.com/search?q=hummingbird+eating+day&rlz=1C1WNCD_enUS895US895&oq=hummingbird+eating+day&aqs=chrome..69i57j33i160l2.10253j1j7&sourceid=chrome&ie=UTF-8

https://www.birdsandblooms.com/birding/attracting-hummingbirds/hummingbirds-can-see-more-colors/?fbclid=IwAR0GzjHM_EbDeo986HXbBiEeGZAX40XFcmXT6lc4MP65fdlM5F6SAUQVWik

http://www.hummingbirdsociety.org/index.php

https://www.hummingbirdspot.com/legends

https://www.kandfoto.com/blog/the-legend-of-the-hummingbird/

https://www.mentalfloss.com/article/62820/9-adorable-facts-about-hummingbirds#:~:text=A%20hummingbird's%20brain%20makes%20up,percent%20of%20our%20body%20weight.)&text=They%20can%20even%20recognize%20humans,to%20refill%20empty%20hummingbird%20feeders.

http://www.native-languages.org/legends-hummingbird.htm

https://www.nytimes.com/2020/06/19/science/hummingbirds-color-vision.html

https://ondessonk.com/heepwah/stories/

https://www.perkypet.com/articles/the-sun-in-disguise

https://www.popsci.com/science/hummingbirds-fast-dives/

https://www.richardalois.com/symbolism/hummingbird-meaning

http://www.rubythroat.org

https://www.sciencealert.com/hummingbirds-can-see-colours-we-can-t-even-imagine-experiment-reveals

https://www.sciencemag.org/news/2020/07/hummingbirds-can-count-their-way-food?utm_source=newsfromscience%3Dflipboard%3Dflipboard2467024

https://www.theastrologyweb.com/spirit-animals/hummingbird-meaning-symbolism

https://theconversation.com/how-colonialism-transformed-foxgloves-and-why-hummingbirds-might-have-had-a-role-158799

https://theconversation.com/most-birds-cant-taste-sugar-heres-why-the-hummingbird-can-31486

https://www.uniguide.com/hummingbird-meaning-symbolism-spirit-animal-guide/

https://www.worldbirds.org/what-does-a-hummingbird-symbolize/

https://www.worldofhummingbirds.com/history.php

SEE HUMMINGBIRDS FLY, SHAKE, DRINK IN AMAZING SLOW MOTION – NATIONAL GEOGRAPHIC
https://www.youtube.com/watch?v=RtUQ_pz5wlo

ABOUT THE AUTHOR

Matthew J. Pallamary's works have been translated into Spanish, Portuguese, Italian, Norwegian, French, and German. His historical novel of first contact between shamans and Jesuits in 18th century South America, titled, **Land Without Evil** received rave reviews along with a San Diego Book Award for mainstream fiction. It was also adapted into a full-length stage and sky show, co-written with and directed by Agent Red and performed by Sky Candy, an Austin Texas aerial group. The making of the show was the subject of a PBS series, Arts in Context episode, which garnered an EMMY nomination.

His nonfiction book, **The Infinity Zone: A Transcendent Approach to Peak Performance** is a collaboration with professional tennis coach Paul Mayberry that offers a fascinating exploration of the phenomenon that occurs at the nexus of perfect form and motion. **The Infinity Zone** took 1st place in the International Book Awards, New Age category and was a finalist in the San Diego Book Awards.

His first book, a short story collection titled **The Small Dark Room Of The Soul** was mentioned in The Year's Best Horror and Fantasy and received praise from Ray Bradbury and has been released as an audio book.

His second collection, ***A Short Walk to the Other Side*** was an Award Winning Finalist in the International Book Awards, an Award Winning Finalist in the USA Best Book Awards, and an Award Winning Finalist in the San Diego Book Awards. It has been released as an audio book.

DreamLand a novel about computer generated dreaming, written with legendary DJ Ken Reeth won first place in the Independent e-Book Award in the Horror/Thriller category and was an Award Winning Finalist in the San Diego Book Awards. It has also been released as an audio book.

It's sequel, ***n0thing*** is titled after the main character, who in the real world is his nephew, an international Counter-Strike gaming champion. After winning what amounts to the Super Bowl of gaming, n0thing and his winning teammates, are recruited as a literal "dream team" whose mission is to go into the nightmares of battle scarred veterans and rescue them from their traumatic memories while becoming ambassadors for a gaming platform that exceeds virtual reality with an experience that pushes the boundaries of reality itself.

Eye of the Predator was an Award Winning Finalist in the Visionary Fiction category of the International Book Awards. ***Eye of the Predator*** is a supernatural thriller about a zoologist who discovers that he can go into the minds of animals.

CyberChrist was an Award Winning Finalist in the Thriller/Adventure category of the International Book Awards. ***CyberChrist*** is the story of a prize winning journalist who receives an email from a man who claims to have discovered immortality by turning off the aging gene in a 15 year old boy with an aging disorder. The forwarded email becomes the basis for an online church built around the boy, calling him CyberChrist. It has also been released as an audio book.

Phantastic Fiction – A Shamanic Approach to Story took first place in the International Book Awards Writing/Publishing category. ***Phantastic Fiction*** is Matt's guide to dramatic writing that grew out of his popular Phantastic Fiction Workshop.

Night Whispers was an Award Winning Finalist in the Horror category of the International Book Awards. Set in the Boston neighborhood of Dorchester, *Night Whispers* is the story of Nick Powers, who loses consciousness after crashing in a stolen car and comes to hearing whispering voices in his mind. When he sees a homeless man arguing with himself, Nick realizes that the whispers in his head are the other side of the argument.

His memoir *Spirit Matters* detailing his journeys to Peru, working with shamanic plant medicines took first place in the San Diego Book Awards Spiritual Book Category, and was an Award-Winning Finalist in the autobiography/memoir category of the National Best Book Awards.

The Center Of The Universe Is Right Between Your Eyes But Home Is Where The Heart Is was an Award Winning Finalist in the International Book Awards. Based on a lifetime of research into shamanism, visionary states, the evolution of written communication and the roots of storytelling, award-winning author, editor, and shamanic explorer Matthew J. Pallamary takes those with open minds courageous enough to question the illusions that most of us think of as real on an expansive journey that pierces the veil of reality itself.

AfterLife: The Adventures of a Lost Soul was inspired by real life events, William Peter Blatty's *The Exorcist*, and the dynamics of demonic possession.

Matt has also produced and directed *The Santa Barbara Writers Conference Scrapbook* documentary film and co-wrote the book of the same title in collaboration with Y. Armando Nieto, and conference founder Mary Conrad.

Death: (A Love Story) a first person narrative spoken by the omniscient voice of Death itself, who says, "I'm here to tell you stories and share some science, history, and myths, all of which are your creations that I want to share to help you understand me more. You have seen me as Satan, Anubis, Mot, Thanatos, God, the Devil, loving, punitive, dark, light – the list goes on and on! It is my sincerest hope that our friendly reintroduction here will change the way you think of me, and maybe in some small way reflect the depth of the love I have for you.

Picaflor is the sequel to ***Spirit Matters***, a San Diego Book Award winner and an Award-Winning Finalist in the National Best Book Awards that chronicles the two decades since of Matthew (Mateo) J. Pallamary's adventures in ***Spirit Matters*** through the mountains, deserts, and jungles of North, Central, and South America pursuing his studies of shamanism and visionary experience working with plant medicines and shamanic plant diets, among them Ayahuasca, Peyote, San Pedro cactus, and many more.

Matt's work has appeared in Oui, New Dimensions, The Iconoclast, Starbright, Infinity, Passport, The Short Story Digest, Redcat, The San Diego Writer's Monthly, Connotations, Phantasm, Essentially You, The Haven Journal, The Hurricanes & Swan Songs Anthology, The Santa Barbara Literary Journal, The Closed Eye Open, The Montecito Journal, and many others. His fiction has been featured in The San Diego Union Tribune which he has also reviewed books for, and his work has been heard on KPBS-FM in San Diego, KUCI FM in Irvine, television Channel Three in Santa Barbara, and The Susan Cameron Block Show in Vancouver. He has been a guest on the following nationally syndicated talk shows; Coast to Coast with George Noory, Paul Rodriguez, In The Light with Michelle Whitedove, Susun Weed, Medicine Woman, Inner Journey with Greg Friedman, and Environmental Directions Radio series. Matt has appeared on the following television shows; Bridging Heaven and Earth, Elyssa's Raw and Wild Food Show, Things That Matter, Literary Gumbo, Indie Authors TV, and ECONEWS. He has also been a frequent guest on numerous podcasts, among them, The Psychedelic Salon, Black Light in the Attic, Third Eye Drops, C-Realm, Psychedelics Today, Voices in the Dark, Adventures Through the Mind, Beyond the Veil, and many others.

Matt received the Man of the Year Award from San Diego Writer's Monthly Magazine and has taught a fiction workshop at the **Southern**

California Writers' Conference in San Diego, Palm Springs, and Los Angeles, and at the **Santa Barbara Writers' Conference** for over thirty years. He has lectured at the Greater Los Angeles Writer's Conference, the Getting It Write conference in Oregon, the Saddleback Writers' Conference, the Rio Grande Writers' Seminar, the National Council of Teachers of English, The San Diego Writer's and Editor's Guild, The San Diego Book Publicists, The Pacific Institute for Professional Writing, The 805 Writers Conference, The College of Central Florida, Yakima Valley College in Washington, The Yakima Public School System, and he has been a panelist at the World Fantasy Convention, Con-Dor, and Coppercon. He is presently Editor in Chief of Mystic Ink Publishing.

Matt was a featured lecturer and performer at the **Mysteries of the Amazon** exhibit at the Appleton Museum in Ocala Florida and The Larson Gallery in Yakima Washington. He frequently visits the mountains, deserts, and jungles of North, Central, and South America pursuing his studies of shamanism.

MATTPALLAMARY.COM

BOOKS BY MATTHEW J. PALLAMARY

THE SMALL DARK ROOM OF THE SOUL

LAND WITHOUT EVIL

SPIRIT MATTERS

DREAMLAND (WITH KEN REETH)

THE INFINITY ZONE (WITH PAUL MAYBERRY)

A SHORT WALK TO THE OTHER SIDE

CYBERCHRIST

EYE OF THE PREDATOR

PHANTASTIC FICTION

NIGHT WHISPERS

THE SANTA BARABARA WRITERS CONFERENCE SCRAPBOOK
(WITH MARY CONRAD & Y. ARMANDO NIETO)

n0THING

AFTERLIFE: THE ADVENTURES OF A LOST SOUL

THE CENTER OF THE UNIVERSE
IS RIGHT BETWEEN YOUR EYES
BUT HOME IS WHERE THE HEART IS

DEATH: (A LOVE STORY)

PICAFLOR